The System is the Boss

How AI, Bad Data, and Broken Metrics Are Reshaping Work

Rad Stephens

For Becky. In a house where every question got a real answer, not a metric or an excuse, I learned what accountability actually sounds like. You gave me that before I ever wrote a word about the absence of it.

And for Amanda, Cortney, Alec, and Emily. Everything I have tried to understand about work, fairness, and what people deserve from the systems they serve ultimately comes back to you. I have always wanted those systems to be better for the people who come after us, and most of all, better for you.

The System is the Boss: How AI, Bad Data, and Broken Metrics Are Reshaping Work

First edition

Published by Rad Stephens Media LLC

ISBN hardcover: 979-8-9965245-2-5

Table of Contents

A Note on Sources and Characters

The authority this book claims is operational. It comes from more than three decades spent inside large organizations: managing teams, running operations, sitting in the rooms this book describes, and making many of the decisions it examines. The argument did not begin as a thesis. It began as an accumulation of experience that eventually became impossible to leave unnamed.

The characters in these pages are composites. Marcus, Denise, Ray, Terrence, and Margaret are not portraits of specific individuals, and none should be read as a disguised account of one person. They are built from patterns I encountered repeatedly across organizations, roles, operating environments, and different periods of my career. Their experiences combine situations, behaviors, conversations, and consequences that occurred in different places and at different times so that the larger structure can be examined without exposing people who never agreed to become characters in a book.

That distinction matters because the purpose of these characters is not to reconstruct individual histories. It is to make systemic behavior visible at the level where people actually experience it. A metric does not feel abstract when someone has to explain it to a team. A routing model is not simply an algorithm when a driver is standing on a street deciding whether to follow what the device says or what eleven years of experience tells him. A staffing recommendation becomes something different when a manager has to look at fourteen people on a floor and reconcile the number on the screen with the work directly in front of him. Marcus, Denise, Ray, Terrence, and Margaret give those pressures a human location, but the patterns they represent extend well beyond any one person or organization.

The situations they move through are drawn from real operational contexts and from conditions I have observed, managed, measured, or worked inside directly. Some events have been combined, timelines compressed, circumstances adjusted, and identifying details changed where necessary. Conversations may represent the substance of several similar exchanges rather than a verbatim transcript of one meeting. The specificity is intentional because systems become difficult to understand when they are described only at the level of theory. The details are there to show how the structure behaves when it reaches the floor, the meeting, the route, the review, and the person whose name eventually appears in the record.

No character in this book represents a single individual, and no scene should be understood as a transcription of one event. That does not make the underlying dynamics fictional. The opposite is true. The composite approach allows recurring patterns to be shown more clearly by separating the argument from the identifying details of any particular workplace or person. What matters here is not whether one manager said one sentence in one building on one Tuesday morning. What matters is that variations of that conversation happened often enough, across enough settings, for the structure behind it to become recognizable.

The technology in these pages is described with the same intent. Internal product names, proprietary terminology, organization-specific labels, and unnecessary technical identifiers have generally been replaced with plain-language descriptions of what the systems actually did. A reader does not need the internal name of a forecasting platform to understand what happens when its output becomes the staffing assumption for the next system in the chain. The important fact is the function the technology performs, the authority its output receives, and what happens when that output encounters conditions the system did not fully capture.

In some places, related technical functions are described together because their combined operational effect matters more than the boundaries between individual products. That should not be read to mean that every capability described existed inside a single application, shared one architecture, or operated as one unified artificial intelligence system. Modern operations are often built from layers of tools, databases, models, dashboards, workflow systems, and automated processes developed at different times for different purposes. They can shape the same decision without technically belonging to the same product. From the perspective of the person doing the work, however, those distinctions can matter far less than the combined environment those systems create.

Terms such as forecasting agent, labor scheduling agent, routing agent, and monitoring agent are therefore functional descriptions. They identify what a system or group of systems is doing inside the operational chain rather than naming a particular commercial or proprietary technology. A forecasting function estimates what work is coming. A labor function translates that expectation into staffing conditions. A routing function determines how work should move. A monitoring function observes the resulting activity and identifies deviations. Whether those capabilities exist as separate applications, integrated services, model-driven workflows, or combinations of older and newer technologies does not change the central issue examined in these pages: one system's output increasingly becomes the condition another system treats as fact.

That distinction becomes especially important as artificial intelligence is added to systems that were already automated. Not every automated decision described in this book was produced by what would now be called AI, and it would be inaccurate to retroactively apply that label to every algorithm, dashboard, optimization model, or rules-based process. The larger pattern developed before the current generation of AI systems arrived. What AI changes is the speed, reach, adaptability, and increasing capacity of these systems to interpret information and initiate action with less direct human involvement. The underlying transfer of authority was already underway.

This book therefore makes a distinction between technical precision and operational truth. Technical precision asks exactly which product produced which output, which model version was running, and where one system formally ended and another began. Those questions matter in engineering, governance, auditing, and implementation. Operational truth asks what happened to the person receiving the combined result. Did the forecast shape the staffing plan before the manager arrived? Did the route become an expectation before the driver saw the street? Did the monitoring system treat the consequences of upstream decisions as evidence about one worker's performance? Those are the questions this book is primarily trying to answer.

That approach is also a choice about where authority lives. Many of the dynamics examined here do not appear in formal reports with the granularity required to understand how they actually operate. A policy can tell you how an exception is supposed to be handled, but it cannot always tell you what happens when the manager knows that using judgment will create a deviation that must later be explained. A dashboard can show a performance trend, but it cannot necessarily show the moment an experienced worker stopped raising concerns because the concerns had repeatedly gone nowhere. An organizational chart can show who has authority on paper while revealing very little about who can actually change a decision after a system has already acted.

Those realities appear in buildings, on routes, in review meetings, in conversations after the official meeting has ended, and in the ordinary decisions people make when no one is documenting the moment as significant. They appear in the pause before someone decides whether to challenge the number on the screen, in the workaround an experienced worker stops reporting, and in the explanation a manager carefully translates into language the system will accept. Much of what matters most about operational life exists in those spaces before it becomes data, if it ever becomes data at all.

That is where this book comes from. Its authority is not the authority of a controlled study claiming that every organization behaves identically, and it is not an argument that every system produces the same consequence in every setting. It is the authority of repeated operational observation: seeing similar structures appear under different names, in different industries, with different technologies and different people occupying the roles. The details change. The recurring relationships between measurement, authority, judgment, accountability, and consequence are what made the pattern difficult to ignore.

The characters are the vehicle for that knowledge, not a substitute for it. They allow the reader to follow the system all the way down to the point where an abstract organizational choice becomes somebody's Tuesday morning. Their names are constructed. The pressures surrounding them are not. The purpose of using composites is to protect the people behind the experiences while preserving enough of the operational truth to examine the system honestly.

The Decision Was Already Made

The decision was already made before anyone walked into the room. Not in whispers, not in politics, not in the usual ways decisions have always been shaped in hallway conversations or pre-meetings. This was different. The recommendation had been generated overnight, processed through a system no one in the room had fully reviewed, and delivered in the morning with the quiet authority of something that did not need debate. No one said that out loud. They did not need to. The presentation moved quickly, a summary slide, a set of projected outcomes, a recommended course of action framed in the language organizations have learned to trust. Data-driven. Risk-adjusted. Aligned with historical performance. The phrasing ended the conversation before it began.

Around the table were people with decades of direct experience, people who had seen variations of this situation before, who understood the conditions in ways no model could fully capture. One of them spoke up, not aggressively, not to challenge the system outright, but to add something it could not see yet. A shift in conditions on the ground. Local knowledge that had not yet crossed any reporting threshold. The response was familiar, professional. Is that reflected in the data? It wasn't. Can you send that through so we can incorporate it into the next review cycle? The next review cycle was weeks away. The decision was being made now. The conversation moved on. This is how everything changed: quietly, incrementally, without anyone marking the moment as a shift.

I remember the exact moment I understood this. I was sitting in a regional leadership meeting, watching a leader walk through a set of recommendations with the measured confidence of someone who had already won the argument before it started. The data was on the screen. The conclusion was embedded in the data. The conversation that followed was not a debate. It was a ceremony. People asked clarifying questions. Nobody asked whether the premise was right. I sat there watching it unfold and realized, with a clarity that felt almost physical, that the authority in that room did not belong to the person at the front of it. It lived somewhere upstream, in the system that had produced the numbers, framed the options, and quietly removed everything else from consideration before we arrived. That realization did not make me angry. It made me want to understand. This book is that understanding.

If you have been in a room where a decision was already made before the meeting started, and where the most professionally sound move was to ask a clarifying question rather than the actual question, this book is for you. Not for you in the sense that it flatters you or argues entirely on your behalf, though it may do some of both. For you in the sense that it names what you have been living in and gives it a structure you can see clearly enough to examine. The people this book is most useful to are not the ones who designed these systems and believe they are working, or the ones who have already left and attributed the experience to a bad organization or a bad boss. It is for the people still inside these environments: managing inside them, executing inside them, absorbing their consequences, and doing all of it without quite having language for what the experience actually is. That is most of the people doing this work. Whether you are on the floor or in the meeting, whether you are Terrence or the person Terrence reports to, if you have ever felt the specific weight of being asked to carry a decision you had no part in making, the chapters that follow are trying to give that weight a name.

Take Terrence. He manages a team of fourteen on a high-volume distribution floor, has done it for six years, and on a Tuesday morning in the middle of a standup he stands in front of his team and walks them through a change in pick rates that does not quite add up when you run the numbers against the current headcount. The room is already tired. This is the third change in two months. The new expectation sounds reasonable in the abstract, the way most expectations do when they are read aloud from a slide, but everyone listening is quietly doing the math, and the math does not work.

One of his team members, a woman named Margaret who has been on that floor longer than Terrence, asks the obvious question: where is the time supposed to come from. It is not a challenge. It is a reasonable question from someone who does the work and wants to understand how to do it correctly.

Terrence pauses, not because he has not thought about the question but because he has, and the honest answer is that the expectation was built at a level where the floor has no visibility, using a model that does not run on the same inputs the floor actually works with. There is no version of that answer he can give Margaret that will satisfy her and still keep the meeting moving. So, he does what the system has trained him to do. He restates the expectation in different words, gestures at the target, and moves on to the number that needs to improve by end of week. The conversation that could have happened does not happen. The question is not answered. The work continues.

The targets Terrence communicated to his team were not set by a senior leader looking at the floor and making a judgment call. They were produced by the company's forecasting and labor optimization system, an AI application that processes transaction volume, historical rates, facility data, and productivity benchmarks and outputs what performance should look like, given all of those inputs, at a level of granularity no human team could replicate manually. It does not appear as AI when it reaches Terrence. It appears as a metric on a dashboard, a rate target in a review document, a number that is simply there, derived, authoritative, and beyond the reach of any conversation happening on the floor. Leaders defer to it because it speaks in the language of data and they have been trained to trust that language more than intuition or testimony from the floor. Managers execute it because that is what the role requires. The people closest to the work adapt to it. They adjust, they absorb the variance, they find workarounds, because adaptation is the only available response when the thing you are adapting to cannot be argued with.

Two words appear throughout this book with a specific meaning, because the language organizations use for these roles is part of what the book examines. When "leader" appears, it refers to the people who make decisions about the system, who set targets, define structures, and determine what gets measured and how. When "manager" appears, it refers to the people who carry those decisions out in real time, who stand in front of rooms like the one Terrence stood in, and who are also called leaders, which is part of the problem. On paper, the language implies distributed authority. In practice, decisions are made at a distance, and execution happens up close, where the consequences land on people like Margaret, who asked a fair question and did not get a real answer. That distinction between deciding and executing, between setting direction and absorbing impact, is the axis this book turns on. The system is designed, not maliciously but structurally, to obscure it.

The argument is simple enough to state and uncomfortable enough to sit with: in most modern organizations, the system is the actual boss. The people above you in the hierarchy still matter. They can protect you or expose you, reward you or sideline you. But the fundamental direction of your work, the range of decisions available to you, the definition of success you are measured against, and the pressure you feel at eleven o'clock on a Wednesday night all originate somewhere the org chart does not show. They come from the system, from the processes, metrics, algorithms, and structures built to create consistency at scale, and that now define, far more than any individual leader ever could, what is possible and what is not. This is not an argument against systems. Systems solved real problems. They removed bias where individual judgment had been weaponized. They created accountability where none existed before. They enabled organizations to grow beyond what any single person could oversee. The achievements are genuine. What I am arguing is that we have not been honest about the trade, that we built environments in which measurement became authority and then pretended that authority still lived in people. The system does not replace people. It replaces the need to trust them.

That cost does not distribute evenly. It concentrates. It falls on the people furthest from the decision, the ones who were never in the room where the model was trusted, the ones who absorb the downstream consequences of choices made with confidence at levels above them. Understanding the system means understanding that distribution. It means naming who pays, and why, and whether anyone in authority is aware enough to care. The decision, in almost every case that follows in this book, was already made before it arrived at the floor. By the time it reached Terrence, by the time it reached Margaret, by the time it reached the customer who started getting packages a day later and never understood why, what looked like an open question was already finished. The system just needed people to carry it out. The people called leaders, more often than not, were the ones doing the carrying.

The Rise of Measurable Everything

The room was loud before the dashboard showed up. Not chaotic. Not unstructured. Just human. People talked over each other. Someone walked in late with information that changed the direction of the conversation. A manager would say, that doesn't feel right, and the room would slow down long enough to figure out why. You could hear the work in those meetings: the friction, the edges, the resistance of reality pushing back against whatever clean story someone had walked in with. That friction was not a problem to be solved. It was the sound of people actually thinking together, and of people who still believed that thinking together was what the meeting was for.

Then the screen came on. I remember the first time I watched a room go quiet because of a dashboard. Not respectful quiet. Not focused quiet. A different kind entirely: the kind where people stop thinking out loud and start checking themselves against what they are seeing. The numbers sat there like they had already settled the argument. Nobody said that out loud. Nobody needed to. A senior leader who had spent twenty years developing the instincts to know when something was wrong on a floor started nodding along with a trend line that pointed in exactly the opposite direction from what he had been telling me privately all week. I watched it happen. He looked at the screen. He looked at his notes. He put his notes away.

We told ourselves this was progress. It was, in part. For the first time, we could see patterns we had been missing for years. Problems surfaced faster. There is a version of this story where measurement saves the day, and it is not fiction. Measurement has prevented failures, exposed blind spots, and forced accountability into places where there used to be none. Anyone who argues otherwise is either naive or selling something. But what we gained in visibility, we quietly traded for something harder to name.

Before everything was measured, judgment had weight. Not because it was always right. It wasn't. But because it had to be defended. A manager who sensed something was off had to say so out loud, in front of people, and stand behind the observation long enough for it to be tested. That exposure was uncomfortable, but it was also honest. The measurement system removed much of that exposure by replacing human testimony with data, and in doing so, it removed something else: the obligation to engage seriously with what you actually saw. Experience was gradually reclassified as anecdote, instinct as bias, and people who had spent twenty or thirty years developing a different kind of knowledge, knowledge earned on the floor, tested through friction, and sharpened by watching enough things go wrong, found themselves holding something the room had stopped accepting as currency.

I have done this more times than I would like to admit. I have stood in a building, watched a process breaking down in real time, and still waited to see whether the numbers would catch up before saying anything. It was not because I distrusted what I saw. I knew what would be asked if I spoke too early: where is it in the data? That question sounds reasonable because sometimes it is reasonable. Evidence matters, and experience can be wrong. But in practice, the question can also become a gate, one that determines which forms of knowledge are allowed into the conversation at all. Once I had been turned back at that gate enough times, I stopped approaching it without the right paperwork.

Here is what most leadership books are too polished to say out loud: we like the system not because it is perfect, but because it protects us. If the numbers are right, the decision is right. If the decision is right, no one is personally responsible for what that decision costs someone they will never meet. The system absorbs the moral weight that used to sit on individuals. That is the deal. And most of us took it without reading the fine print.

Over time, people stop reacting to the work itself and start reacting to how the work will be measured. They learn the thresholds, the triggers, the signals that draw attention from above. Behavior shifts, not through command but through calibration. Like a musician adjusting to a metronome, the work begins to sync to the beat of the metric. But the metronome only captures part of the song. What looks like performance is often adaptation. What looks like improvement is often people getting better at satisfying the measurement, not at improving the underlying reality the measurement was built to represent. This is not manipulation. It is survival.

Measurement stops being a tool and becomes a lens. Then a filter. Then, finally, the authority. No one stands up and declares it. It happens the way the ground moves without anyone noticing. Slow, invisible, layer by layer, until one day the landscape is different and no one can point to when it changed. Geologists call it creep, movement so gradual it goes unrecognized while it's happening, only becoming visible after it's already too late to question how it started. The mountain looks stable. The instruments say nothing alarming. Then one morning it isn't stable, and there is no honest way to say when that became true.

I walk into a meeting. The numbers are already on the screen. Before anyone says a word, before the first slide advances, I already know how the conversation is going to end. The question nobody asks, the question this book is going to keep asking, is whether anyone in that room is still actually deciding anything.

When Data Became Authority

The spreadsheet was already open when we joined the call. Tabs across the bottom, color-coded columns, a summary page at the front that had reduced thousands of decisions into eleven numbers. Someone shared their screen and began walking through it, cell by cell, the way a pilot reads instruments: with a calm that assumes the instruments are right. No one interrupted. No one asked where the assumptions came from. The model was treated like weather: something to navigate, not something to question.

Halfway through, a leader paused on a single line and said, if this holds, we need to make the change this week. That one output would reroute three hundred people, shift capacity across two regions, and tighten timelines in ways that would ripple through the network by morning. I looked at the number. I knew that operation. I had been on that floor six days earlier, and what I had seen there did not match what the screen was showing. The divergence was small. I had been in enough operations to know that small sometimes means nothing has broken yet.

I almost said something. I can still feel the moment. The breath I took, the calculation I ran in the space of two seconds. I had no data to counter it with. What I had was pattern recognition built over years, and in that room, pattern recognition was not an accepted form of payment. Do you have numbers that show something different? The question would have been polite. It would have also been a verdict. So, I stayed quiet, and the decision moved forward, and within four days the friction showed up exactly where I thought it would. Enough to make the work harder for people who were not in that meeting and would never know their difficulty had been predictable.

Data does not start as authority. It earns it. Because it is consistent when people are not, because it can be shared across time zones at two in the morning, because it arrives without ego or emotion attached to it. In a world full of competing opinions, data feels like relief: something you can stand on without defending yourself. Over time, that feeling becomes belief. And belief, unexamined, becomes assumption. We stop asking whether the data is complete and start assuming it is. The conversation narrows. It becomes faster, more efficient, and thinner.

Data doesn't lie. It just doesn't tell the whole truth. And in the space between those two statements, organizations make their most consequential mistakes.

To be fair, because this story will not allow for a clean villain, sometimes the system is exactly right. I have seen data cut through noise that would have derailed a decision and exposed bias so embedded that it had become invisible to the people inside it. Data protects people from leaders who trust their instincts a little too much. I have been that leader. The mechanism that constrains me is the same one that has, on other days, saved me from myself. That is what makes it so difficult to confront honestly.

When data becomes authority, it does not just inform decisions. It defines what a decision is allowed to be. Instead of asking what is the right thing to do, people start asking what the data supports. That sounds responsible. It is also restrictive, because data is always a map drawn from the past, even when we are pretending it is predicting the future. Maps leave things out. They cannot show what it feels like to work in a building at hour eleven of a difficult shift. They cannot capture the hesitation in a room when people know something is wrong but cannot yet prove it.

I once had a senior leader tell me, with complete seriousness, if it's not in the data, it's not real. I understood what he meant. I also understood that the most important signals in any operation show up first in conversations, in hesitation, in the way people start working around a process before anyone admits the process is broken. By the time those signals are clean enough to measure, the damage has usually already started. You cannot put hesitation in a spreadsheet. You cannot graph the moment a team stops raising concerns because raising concerns has stopped working.

So those signals get discounted, not intentionally but structurally. The system rewards what can be proven. Over time, people internalize that, and they stop surfacing what they cannot yet prove. And then something quiet but profound happens: the version of reality the organization is willing to acknowledge shrinks, not because reality changed, but because what the organization is willing to see did.

I still think about that conversation. Nothing about it felt like a failure. The numbers were solid. The decision made sense on paper. If you had reviewed the transcript, you would have found nothing alarming. What you would not have found is the breath I took before I chose silence. That is the moment data stops being a tool and becomes authority: when the cost of speaking without proof exceeds the cost of knowing something and saying nothing. Once authority shifts that way, it is very hard to move back.

The data was only the beginning of that shift. What followed was quieter and more consequential. Organizations that had learned to trust data as a proxy for reality began developing systems that used that accumulated data not just to describe what had happened, but to project what would happen next. It started modestly: trend lines, seasonality models, capacity projections. The tools were limited, the outputs clearly probabilistic, and most of the people using them understood that. Then the tools became more sophisticated, and the understanding did not keep pace. The confidence remained, and it transferred, from data about the past to predictions about the future, with almost no pause to examine whether the confidence was still warranted.

That transfer has not been named clearly inside most organizations, because the output looks the same. A number on a screen. A recommendation attached to a projection. A course of action presented as what the data supports. It all carries the same surface authority, even though the underlying mechanism has changed.

The difference is structural. Historical data is a record of what has already happened, even if it is incomplete. A predictive model is an argument about what will happen next, shaped by assumptions most people in the room cannot see and often do not know to look for.

In large organizations, that argument is no longer abstract. It has been embedded directly into the machinery through which work is assigned, measured, evaluated, and corrected. Forecasts determine how many people are scheduled before anyone sees the conditions of the day. Models distribute workload, establish rate expectations, identify what should be achievable, and define what good performance is supposed to look like before the first person begins the work. Risk scores determine which conditions deserve attention, while performance systems compare what actually happened against what the model predicted should have happened and convert the distance between those two numbers into a finding.

That sequence matters because prediction does not remain prediction for very long. A forecast begins as an estimate, but once the staffing system acts on it, the estimate becomes headcount. Once the headcount reaches the floor, it becomes the operating condition people have to work inside. The productivity model then measures those people against expectations that may have been influenced by the same assumptions that shaped the staffing decision in the first place. What began as an argument about what might happen becomes the environment within which performance is later judged.

The transformation is easy to miss because each step appears reasonable on its own. Of course, the organization should forecast demand. Of course, staffing should reflect expected volume, and of course managers need standards against which to evaluate performance. None of those decisions is inherently problematic. The problem appears when the original uncertainty disappears as the output moves downstream, when the forecast stops being treated as a forecast and begins functioning as a fact that every subsequent system assumes has already been settled.

By the time the number reaches someone like Terrence, very little of that uncertainty remains visible. He does not receive a probability distribution, a list of assumptions, or a careful explanation of the conditions under which the recommendation is most likely to be wrong. He receives a target. The target appears on a dashboard beside actual performance, and the difference between the two becomes something he is expected to explain. The model was allowed uncertainty while producing the expectation, but the manager receives very little uncertainty when being held accountable for meeting it.

That is one of the quiet ways data becomes authority. The system is permitted to make an estimate, but the person receiving the estimate is expected to treat it as an obligation. If the estimate is wrong, the first visible evidence of that error may not appear as a model failure at all. It may appear as a productivity gap, a missed rate, a staffing problem, a route variance, or a manager whose team did not deliver what the system predicted should have been possible. The uncertainty travels downward and gradually changes its name until someone at the point of execution is asked to account for it.

What looks like a recommendation is therefore often closer to a constraint. The system may technically leave room for human judgment, but the surrounding organization determines how expensive that judgment will be to exercise. A manager can challenge the staffing recommendation, but the existing staffing plan is already approved. A driver can alter the route, but the original route remains the benchmark against which the deviation will be measured. A leader can question the projection, but the planning process may already have organized resources around it. The system does not need to prohibit disagreement when the cost of disagreement has already been built into the workflow.

The authority becomes stronger because the output presents itself in the language of neutrality. It arrives as something derived from data rather than something constructed through assumptions, weighting choices, historical patterns, thresholds, exclusions, and definitions of success. Those choices may all be reasonable, but they are still choices. Once they disappear behind an interface, the resulting number begins to look less like an argument and more like an observation about reality. The people receiving it see the conclusion without seeing enough of the reasoning to engage with the conclusion on equal terms.

That difference changes the nature of disagreement. If a person makes a recommendation, you can ask what they considered, what they left out, why they weighted one factor more heavily than another, and what might cause them to reconsider. The reasoning has an owner, and even when you disagree with that owner, there is a human path back through the decision. When a system makes the recommendation, the logic may exist somewhere in the model, configuration, data pipeline, or methodology, but it is often inaccessible to the people being asked to act on the result. They are presented with the answer without being given an equally usable way to interrogate the argument that produced it.

Records behave differently. A record tells you what happened, or at least what the system captured about what happened. You can point to an event, examine the surrounding circumstances, compare accounts, correct missing information, and argue about what the evidence means. A prediction wrapped in institutional authority is harder to challenge because the event it describes may not have happened yet. The person raising the objection is therefore arguing from experience, context, and judgment against something that arrives with statistical weight, organizational sponsorship, and the apparent precision of a number on a screen.

That imbalance becomes especially powerful when the system producing the recommendation is also connected to the system measuring whether the recommendation was followed. The model says the work should require fourteen people, the operation receives fourteen people, and the performance system later asks why fourteen people failed to produce the expected result. The organization can begin evaluating reality against the model instead of evaluating the model against reality. When that happens, a gap between prediction and outcome naturally becomes evidence about execution rather than evidence that the prediction itself may deserve another look.

I have watched that reversal happen without anyone consciously choosing it. The model establishes what should have happened, the actual operation produces something different, and the conversation begins with the variance. Why did the building miss? Why was the route late? Why was productivity below plan? Those are legitimate questions, but they contain an assumption that is rarely stated aloud: that the predicted outcome deserves to remain the fixed point and reality is the thing that must explain why it moved away.

Once that assumption settles into the organization, the system has acquired authority in the deepest sense of the word. It is no longer simply helping people understand reality. It is defining the version of reality against which people must defend themselves. The numbers appear precise, the recommendations appear objective, and the model's conclusion arrives with the accumulated credibility of the organization that selected, tested, approved, and embedded it into the work. The person challenging it arrives with something far less protected: what they saw, what they know, and what the conditions in front of them appear to require.

The alternative, then, is no longer ordinary disagreement. It is deviation from what the system has already determined should happen. That distinction explains why experienced people can see something wrong and still remain silent, why managers can know a target does not fit the floor and still communicate it, and why leaders can recognize uncertainty in a recommendation while allowing the decision to proceed. Data became authority not when organizations began using more of it, but when disagreeing with its conclusions started requiring more evidence, more explanation, and more personal exposure than accepting them.

When Systems Began Talking Without Us

The first change was not dramatic enough to announce itself. No one walked into a meeting and said the systems were now shaping the operation before the people arrived. No one described it as a transfer of authority. It entered through ordinary improvements, the kind every organization learns to accept because each one solves a problem that is real.

A forecasting tool became more accurate. A scheduling system became more responsive. A routing platform became more efficient. A monitoring dashboard became faster at surfacing exceptions. Each system, viewed by itself, made sense. Each one had a purpose, a business case, and a metric that justified its existence. The mistake was thinking they would remain separate.

They did not have to be formally connected to begin affecting one another. One system did not need to send instructions directly to another system for the operation to begin changing around their combined output. The forecast shaped the staffing plan. The staffing plan shaped the conditions on the floor. The floor conditions shaped route timing, completion rates, exception volume, and performance flags. The monitoring system then read those outcomes as if they were independent facts about the work, not consequences of choices other systems had already made.

This is how systems began talking without us. Not through conversation. Not through coordination. Through consequence. A number produced upstream became a condition downstream. A staffing adjustment became a throughput issue. A routing decision became a performance deviation. A performance deviation became a coaching conversation. By the time a manager or worker encountered the issue, the issue had already traveled through several systems, each one treating the previous output as reality.

No single system understood the whole chain. No single person owned it. But the people closest to the work experienced the whole thing as pressure. That is one of the reasons this change was so difficult to see while it was happening. Each part looked narrow, useful, and defensible. The whole felt different, but the whole had no name.

I remember sitting in a planning conversation where the forecast on the screen looked precise enough to end the debate. The volume projection was clean, the labor recommendation followed from it, and the staffing plan looked reasonable if you accepted the assumptions that produced it. The problem was that the assumptions were already stale. The building had changed. The mix of experienced and newer workers had changed. The timing of the work had changed in ways that had not yet made their way into the data. But the forecast had already moved, and once it moved, everything else started moving behind it.

The staffing system did not know the forecast was incomplete. It took the number it was given and built a plan around it, which is what it was designed to do. The plan then reached the floor as a schedule, a headcount, and a set of expectations that carried the authority of a finished decision. The people receiving it did not experience it as a prediction. They experienced it as the day they had to run.

That difference matters. A prediction can still be questioned. A schedule has already acted. Once people have been called in, released early, assigned to stations, or told what volume they are expected to process, the organization is no longer discussing what might happen. It is managing the consequences of what the system already decided was likely enough to build around.

This was not artificial intelligence in the way people usually picture it. There was no single intelligent machine making a grand decision. There were separate systems, each built to solve a narrow problem, each optimized for a different part of the operation. One system looked at volume. Another looked at labor. Another looked at routing. Another watched for deviations. None of them needed to understand the others for all of them to shape the same day.

That made the structure harder to see. If one person had made all of those decisions, the chain would have been visible. You could ask why the staffing plan did not match the floor conditions. You could ask why the route ignored what experienced workers or drivers already knew. You could ask why the monitoring system treated a downstream deviation as a local failure. But when those decisions were distributed across systems, the chain disappeared into the normal flow of work.

Everyone could point to the part they were responsible for and be correct. The forecast was based on the available data. The staffing plan followed the forecast. The route followed the model. The flag followed the threshold. The review followed the flag. Each step was defensible in isolation, which is one reason the whole thing became so difficult to challenge.

This is one of the quietest ways accountability disappears. Not because people stop caring and not because anyone is trying to hide from responsibility, but because each system creates only part of the condition, and each person sees only the part closest to their role. The manager sees the staffing gap. The driver sees the route. The worker sees the rate target. The senior leader sees the dashboard. Nobody is positioned to see the full sequence from assumption to consequence.

That does not mean no one knows something is wrong. Usually, someone does. Often, several people do. The floor manager knows the staffing plan is too tight before the numbers show strain. The experienced worker knows the rate target assumes a cleaner process than the one actually running. The driver knows the routing sequence will fail in a place the map treats as ordinary. Their knowledge is specific, current, and useful. It is also late to the system, because the system has already acted by the time they are asked to respond.

The formal channels do not solve this problem because they usually operate after the fact. A note can be entered. A concern can be raised. A deviation can be explained. But the mechanism for surfacing knowledge and the mechanism for acting on knowledge are not running on the same clock. By the time the concern reaches the level where it might change the model, the model has already changed the day.

The lag teaches people that speaking is not the same as being heard, and being heard is not the same as affecting the decision. After enough cycles, people adjust. They stop entering the note because the note becomes history before anyone acts on it. They stop raising the concern because the concern routes to a review process that will finish after the work has already absorbed the cost. They keep the knowledge, but they stop expecting the structure to use it.

This is where the loss of discretion begins before the visible moment of decision. By the time Marcus is standing there with the package that will not scan, the system has already narrowed the world around him. The route was built before he saw it. The timing expectation was calculated before the road conditions appeared. The monitoring logic was waiting before the exception occurred. The question he asks in that moment, what do you want me to do, sounds like a request for a manager's judgment. But the answer has already been shaped by systems that arrived earlier than either of them.

A manager can still make a call in that moment. That is important to say clearly, because people inside these systems are not powerless. But the call no longer happens in open space. It happens inside a chain of prior decisions that have already defined what will count as success, what will count as deviation, and what will need to be explained later. The manager is not choosing between the customer and the scan in a neutral environment. The manager is choosing while already knowing how the system will record the choice.

That is why the system does not need to issue a direct command. It only needs to make one choice more legible than the other. The scan is legible. The customer's experience is harder to see. Compliance is legible. Judgment is harder to defend. A clean record is legible. A good exception handled well may disappear entirely unless something goes wrong. People learn this difference quickly, because their jobs teach it to them every day.

The same structure applies beyond the floor. A staffing model can be wrong in a way a manager can see but cannot prove in the moment. A forecast can miss a local shift that has not yet entered the data. A route can ignore knowledge that lives in a driver's memory rather than in a system field. A monitoring dashboard can attach a name to an outcome without understanding how several upstream systems helped create it. None of these failures has to be dramatic. Most of them are ordinary, which is why they are so durable.

What makes this more dangerous is that the systems appear calmer than the people. They do not sound frustrated. They do not hesitate. They do not carry the tension of the floor into the room. They produce numbers, sequences, recommendations, flags, and reports with the same tone regardless of how incomplete the underlying picture may be. That calmness becomes authority. The person closest to the work arrives with context. The system arrives with output. In most rooms, output wins.

I have trusted that output when I should have slowed down. I have accepted the forecast because questioning it would have required more time, more evidence, and more exposure than the room was prepared to absorb. I have watched a recommendation move forward because each part of it made sense, even though the whole thing rested on a picture of the operation that people on the ground would have recognized as incomplete. At the time, those decisions felt practical. Looking back, practicality was often the name I gave to moving with a system that had already made resistance expensive.

This is not an argument that the systems were useless or wrong by default. Many of them solved real problems. They reduced manual work, caught errors faster, improved consistency, and gave organizations visibility they had never had before. The issue is not that they existed. The issue is that they began shaping one another's reality without anyone being responsible for the combined effect.

That combined effect is the missing subject in most organizational conversations about technology. The conversation usually stays with the individual tool. Did the forecasting tool improve accuracy. Did the scheduling system reduce overstaffing. Did the routing platform lower drive time. Did the monitoring dashboard surface deviations faster. Those are fair questions, but they are incomplete. The more important question is what all of those systems produce together when they act on the same people, inside the same operation, at the same time.

The answer is not a single system in the traditional sense. It is an ecosystem. Separate tools begin functioning as one environment because every output becomes an input somewhere else. The system does not have to be conscious to behave as if it has momentum. It does not have to be coordinated to produce coordinated pressure. It only has to keep moving, with each part treating the last part's output as settled enough to act on.

Later, as artificial intelligence becomes more deeply embedded in these workflows, this condition becomes easier to name. It begins to look like a hybrid agent ecosystem: forecasting agents, labor agents, routing agents, monitoring agents, each acting separately and shaping the same operation through the consequences they leave behind. But the pattern began before the language arrived. Long before anyone called it an agent ecosystem, the work was already being shaped by systems that did not wait for people to understand the whole before acting on their part of it.

That is why the next loss was not simply discretion. It was the confidence that discretion still had a place to stand. The people closest to the work could still see what the system missed. They could still make judgments. They could still recognize the gap between the rule and the purpose the rule was supposed to serve. But increasingly, they were making those judgments after the systems had already moved, inside conditions the systems had already created, with records the systems were already building around whatever they chose to do next.

By the time the package would not scan, the question was no longer just whether Marcus should take it. The question was what kind of organization had been built around that moment, and why the right answer had become harder to give than it should have been.

The Death of Discretion

The message came through as a reminder, not a directive. Make sure the scan happens before departure. No exceptions. It landed in the middle of a busy shift, and the logic behind it was clean: scan ensures visibility, visibility ensures accountability, accountability ensures performance. Nobody argued with it. It aligned perfectly with how the system defined success.

An hour later, a driver named Marcus was standing in front of me holding a package that would not scan. The barcode was damaged. Not missing, just damaged enough that the system would not recognize it. The clock was moving. His route was already tight. He had been driving for this company for eleven years and had handled a thousand situations more complicated than this one. He looked at me and said, what do you want me to do?

That question, simple, direct, asked by a man who already knew the answer, is the question this entire chapter is about. Because the answer used to be obvious. The package goes. The customer matters more than the metric. Use your judgment, that is why we hired you. But I stood there and did the math the system had taught me to do. If Marcus bypassed the scan, the deviation would be logged. Logged deviations accumulate. Accumulated deviations become a pattern. Patterns become conversations I did not want to have. So instead of answering the question Marcus was actually asking, I told him to try scanning it again. He looked at me the way experienced people look at decisions they recognize as wrong. He did not say anything. He tried it again. It did not scan. The package went out late. The customer experienced delay. The metric showed compliance. The system recorded success.

That was not the first time Marcus had adapted. Three weeks before that conversation, his route had been rebalanced as part of a depot restructuring, and the new sequence looked logical on the routing map, the way most things look logical when they are built at a distance from the conditions they describe. It was not logical to anyone who had driven those roads in October, when the school zones backed up and the afternoon light cut visibility on the crosswalk. Marcus had flagged it in the digital notes field on his scanner, with specifics: the sequence, the schools, the crossing, the time window. He had done it three times across two different routes. Nobody had followed up. So he built his own sequence, kept it in his head, and when the system's route populated on his scanner each morning he followed it on paper and drove what he knew. Two additional stops in a different order, fifteen minutes added, everything delivered on time by a method the system did not understand. His numbers showed higher completion time than the model expected. He had been flagged twice in the last quarter for route variance. He stopped entering alternate sequences in the notes field after the second flag. He just drove his route and let the system record something that was not quite what was happening.

There was another habit Marcus had developed in his third week on the job, after he had delivered medication to the wrong door and watched the system log it as delivered and consider the matter closed. He had started doing a second check on every package before he left the vehicle: open the tote, confirm the address label matched the package label, reseal if needed. Forty-five seconds, every stop. It was never required. It was never counted. It was the kind of thing experienced drivers develop when they have been burned by the gap between what the system confirms and what is actually true. When the throughput pressure increased and his stops-per-hour metric started being reviewed monthly instead of quarterly, he stopped. Not all at once. He stopped on the packages that looked clean. Then on the packages that looked mostly clean. Then the check just was not part of his route anymore, and he could not have named the week it had gone, only that it had.

By November, Marcus had found the rhythm that kept him inside the acceptable band. He skipped the lobby confirmation on units with accessible mailboxes. He stopped waiting for elevators above the third floor and left packages at the mail rooms instead. For signatures, he knocked twice and moved to the next stop; if no one answered by the time he circled back, he logged the return code. His numbers looked better. His completion rate had improved four points. The dashboard registered this as performance improvement. What the dashboard did not register was what four of his regular customers had started to notice: packages marked as delivered appearing in wrong locations, return codes accumulating for people who had been home but had not heard a single knock before he was already gone. Those customers had not complained yet. Most of them would not. They would just stop expecting much, which is the kind of damage that never generates a ticket, that lives entirely below the threshold of what the system is designed to see, that spreads quietly through the experience of people whose expectations the organization has trained itself not to ask about.

What discretion produces is the ability to tell the difference between following a rule and honoring the purpose the rule was built to serve. What the system produces is an environment where that distinction carries a personal cost the system will not absorb.

There was a time when handling exceptions was the core value a leader added. The system managed the standard case. The leader managed the edges, the moments where something did not fit cleanly into a process and someone had to decide what mattered more. That kind of leadership required trust, and it required accepting that two people might make different calls in the same situation and both might be right. The system does not tolerate that variation, because consistency is easier to measure than judgment. And what the system cannot measure, it eventually eliminates.

The argument against discretion is not trivial, and I will not pretend otherwise. Discretion has been used to protect poor performers. It has created uneven standards that undermined trust across teams and shielded bad decisions behind the word judgment. There is a real reason organizations moved toward systems, and that reason was not paranoia. It was accumulated evidence of what happens when individuals with authority are not accountable to anything beyond their own instincts. I have seen that failure mode too. It is real.

But in solving for inconsistency, something important was lost. The system does not distinguish between a thoughtful deviation and a careless one. Both register as the same kind of failure. That flattening effect changes behavior in ways that are invisible to the system, because the system is not designed to see the difference between someone who followed a rule because it was right and someone who followed it because the alternative was too expensive. The outputs look identical. The organizations look identical. The people inside them are not.

People do not lose discretion all at once. They give it up incrementally, in moments like the one with Marcus, in moments like the ones Marcus was living before anyone asked him a direct question. Small, rational, self-protective decisions that accumulate until the question what do you want me to do stops getting asked. Because eventually, the system has already answered it. And everyone in the building knows it. What I lost, in the quiet accumulation of those moments, was not the decision. I lost the person who had been capable of making it.

Incentives Over Intentions

The meeting where the new Sort Completion metric, a measure of how quickly packages cleared the sort floor each shift, was introduced felt optimistic. The leader presenting it was someone I respected: someone who had spent time on the floor, who understood the work, who had genuine intentions about what improvement should look like. The metric was designed to increase speed. The targets were ambitious but achievable. The logic was sound: what gets measured gets attention, and this was worth attending to. I remember thinking it might actually work.

It worked. And then it became something else. Within a quarter, the numbers were moving in the right direction. Reports showed improvement. The metric was cited in presentations as evidence that the change had been effective. At the same time, the work felt different in ways that were not showing up anywhere. Conversations that used to happen, the informal catches, the double-checks, the small acts of colleagues covering for each other, were getting shorter or disappearing. Nobody had told anyone to skip those things. Nobody had said accuracy matters less than speed. But the signal was clear, the way a thermostat is clear: the environment had been set, and behavior adjusted to meet it. The system was working exactly as designed. The intention behind the system was already a memory.

When an incentive replaces the intention that created it, the organization has technically succeeded and actually failed. The gap between those two things is usually invisible until it isn't.

I have pushed teams to hit a number because I knew that number mattered to the system, even when I could feel the strain it was putting on everything the number didn't measure. I told myself it was temporary. I told myself we would correct for it later. That story is easy to tell in the short term and expensive in the long term, and I have told it more often than

I should have. The correction sometimes came. When it didn't, the metric had already moved, and we were chasing the next one, and whatever we had left behind had been normalized.

Jazz musicians understand something most systems ignore. In a good solo, the notes you leave out matter as much as the ones you play. Space carries meaning. Incentive systems are not built to see that. They measure what happens, not what was avoided, and over time that shapes behavior in ways that no one names directly but everyone on the floor eventually learns to feel. They cannot reward the corner that goes uncut, the decision to slow down and prevent an error that no one would have traced back to its source, the twenty-minute conversation that saves three weeks downstream. Those actions do not translate into metrics. Over time, they stop being performed.

What gets lost is not quality, at first. It is the tissue connecting everything: the informal attention, the extra minutes, the investment that comes from people who feel like their judgment matters. That tissue holds organizations together in ways that only become visible when it is gone, when the numbers are still good and something feels hollow, when performance is technically meeting expectations and everyone in the building knows the place is running on diminishing reserves.

The tension this creates cannot be cleanly resolved. Leaders are asked to drive results the system defines while preserving qualities the system cannot see. Hit the number. Maintain the culture. Move with urgency. Stay thoughtful. Those expectations are not unreasonable individually. Together, in practice, they are often incompatible, and most organizations do not acknowledge that incompatibility directly, which means the people inside them are left to carry it alone.

The metric improves. The intention fades. The organization reports success and calls it progress. The people who built the intention in the first place watch it go without having a word for what they are watching.

The Illusion of Control

The weekly review started the same way it always did. Dashboard on the screen, green across most categories, a few yellows already explained away in conversations that had happened before any of us arrived. A senior leader scanned the summary and said, looks like we're in a good place. The room proceeded on that basis.

I was sitting in that room knowing that the picture was incomplete. Not wrong. Incomplete. There were things building on the ground that had not yet crossed a threshold the system was configured to detect. I had been in those buildings. I had talked to the people doing the work. What they were describing to me privately was not visible on any of the screens in that room. I knew what I knew. I also knew that what I knew, in that room, in that format, did not count. The dashboard said we were in a good place. My conversations with frontline managers said something different. One of those sources had a chart. The other one didn't. The meeting continued.

A stable dashboard is not the same as a stable operation. The system only knows what it is measuring, and it is only measuring what someone decided to measure, and those decisions were made in a different moment by people who were not thinking about this specific situation. That gap, between what the system tracks and what is actually happening, is not a malfunction. It is a design limitation that gets treated as proof of functionality.

Control, as the system defines it, is reactive. It is calibrated to detect problems after they cross a threshold, not before. Real control, the kind that prevents collapse rather than responds to it, requires attention to things that do not yet have a number attached to them, and a willingness to act before the system confirms that action is necessary. That kind of control feels uncertain. It cannot be shown on a slide. It does not produce the clean, defensible record of decision-making that organizations increasingly require. So, it fades.

I have caught myself doing this: using the stability of the dashboard as permission not to look harder. Not out of laziness, but out of the accumulated logic of operating in an environment where looking harder than the system requires is treated as noise rather than diligence. The system does not reward leaders who were right before the data confirmed it. It rewards leaders who act on data. And over time, I learned to wait for it.

There was a quarter where everything looked fine on paper. Metrics held, targets were met, reports were clean. On the ground, conversations were shorter. Small issues were taking longer to resolve. Nothing had broken, but something was shifting in a way the system had not yet registered. We waited for the data to reflect it. By the time it did, the correction required was three times what it would have been if we had acted on what I knew and couldn't yet prove.

The deeper problem is that the system trains leaders to treat silence as reassurance. The absence of an alarm becomes evidence that nothing is wrong. But silence is what the system produces when nothing has crossed a threshold, which is entirely different from nothing being wrong. The most dangerous place in any organization is not the team in crisis. The most dangerous place is the team that appears stable because the metrics haven't caught up yet. That place has no alarm. That place looks green. What the system produces in that moment is not control. It is verification, and verification is always, by definition, too late.

The newer versions of those dashboards do not stop at what happened. They show what the system expects will happen: projected numbers, predicted gaps, anticipated risks surfaced before they have materialized. In theory, this extends the window for real control. In practice, it extends the window for the illusion of it. The projection is generated from historical data, the same quarterly patterns, the same operational assumptions the organization has always made, the same variables the system was configured to track. When conditions are stable and the present closely resembles the past, the projections can be reliable enough to be genuinely useful. When conditions are shifting, when something new is building on the ground of the kind that shows up first in conversations rather than dashboards, the prediction model does not register that. It extends the past forward with the same confidence it applies to anything else. The screen still looks green. The projected line still runs upward. And if the baseline was already incomplete, every projection built on top of it inherits every gap and carries them forward, quietly compounded.

I have sat in those rooms with an incomplete baseline visible to me and let the projection run. Not once. The first time I told myself it was tactical, that raising it would cost more than I could afford in that environment at that moment. By the second or third time, I had stopped framing it as a decision at all. It had become the default. The screen looked reasonable, the direction was already in motion, and I had learned what those rooms wanted from me. What they wanted was not what I knew. What I knew stayed where it was, and the projection moved forward, and eventually someone at the floor level absorbed the distance between what the model had predicted and what the conditions had actually required. That distance never appeared in any root cause analysis with my name on it.

The Dashboard Is Watching

The first time someone said it out loud, it got a laugh. "Careful, the dashboard is watching." People appreciated the release. But the joke carried more truth than the laughter acknowledged, and everyone in the room understood what it meant. Not that the dashboard was literally watching, but that everything was being recorded in ways that would outlast the moment. That record would be used to form conclusions about people long after they had forgotten the specific decision that created it. The visibility was total. Every action had a timestamp. Every deviation had a record. The system could reconstruct what happened with a precision that would have been impossible a decade earlier.

At first, that visibility feels external, something you check after the fact, something that reflects what has already occurred. Over time, the relationship inverts. The dashboard stops being a mirror and becomes a presence. Not physically in the room, but in the way people think about the room. Decisions begin to pass through a second layer of consideration, not just what is right or necessary, but how it will register. I watched this happen to myself gradually enough that I did not notice it was happening until the habit was already formed. I would frame a decision in terms that aligned with the metric even when the underlying reasoning was more complex. Not to deceive. To survive. The environment had been configured to reward a certain kind of legibility, and I had learned to be legible.

People do not become less capable when they are watched. They become more careful. And careful, in this context, is not always the same as good.

The psychological change is precise. People begin to optimize not just for the outcome, but for the appearance of the outcome. In many situations, those are identical. In the situations where they diverge, where doing the right thing would look wrong on a dashboard, or where the right-looking thing would require leaving the right thing undone, the gap between them becomes the defining choice of that moment. Organizations that cannot measure that gap cannot see how often it is resolved in favor of appearance.

What begins to disappear is the willingness to act without knowing exactly how the action will be recorded. Spontaneity tightens. Experimentation becomes a calculated risk rather than a natural impulse. The system does not punish failure directly, but it makes failure permanently visible and easily compared. Over time, that visibility becomes more powerful than any rule. Nobody tells people to stop trying new things. The system just makes the cost of trying and failing sufficiently legible that people stop doing it without needing to be told.

The same dynamic exists in places where the stakes are far more immediate. There is something surgeons understand about operating under observation that applies here. A surgeon can perform the same procedure hundreds of times until the movements become almost instinctive. The hand knows where to go. The eyes know what to look for. The mind holds the whole field at once, the tissue, the bleeding, the instrument, the next decision. But place that same surgeon under formal observation, with a senior physician standing behind them, a resident watching from the corner, and a quality review waiting afterward, and the work changes.

The incision may be the same. The anatomy may be the same. The patient may need the same judgment. But now the surgeon is carrying another task inside the procedure. They are not only thinking about what the body is telling them. They are thinking about how their choices will be read. Was that pause too long? Did that adjustment look uncertain? Should they explain what they are doing before anyone questions it? The operation becomes divided between the patient on the table and the impression forming in the room.

Awareness of being watched does not just change behavior at the margins. It changes the cognitive load of the work, redirecting attention that would otherwise go toward the problem and pulling it toward the impression being made. The surgeon begins to manage the room while managing the body. The more automatic that self-monitoring becomes, the less fully present the surgeon is inside the decision.

The dashboard does something similar. It teaches people how to perform for the dashboard without ever saying a word. It does not need to shout instructions or issue direct commands. It changes where attention goes. It teaches people to anticipate judgment before judgment arrives. That is the more sophisticated kind of authority, not the kind that commands, but the kind that configures.

The system does not need to punish bad behavior. It just needs to make good behavior and visible behavior feel like the same thing. Once people stop being able to tell the difference, the work is done.

AI Can Snitch on You

Marcus did not remember the stop as anything unusual. The route had put him behind early, the school-zone sequence had cost him time exactly where he expected it would, and later in the day he took one stop slightly out of order because doing so made sense from the street. He knew the building, knew the traffic pattern, and knew that following the sequence exactly would cost more time than changing it. He made the adjustment, completed the stop, and kept moving. Nothing about the decision felt significant enough to remember because experienced people make dozens of choices like that during a shift, most of them small corrections between what the system expects and what the conditions actually require.

Nothing happened that day. No manager called him, no coworker complained, and nobody standing nearby decided the change was serious enough to mention. Marcus finished the route and went home with no reason to believe the moment would survive beyond the shift. There had been no confrontation, no obvious failure, and no person who appeared to care that he had moved one stop ahead of another. As far as Marcus knew, the decision had ended when the delivery was completed and he moved to the next address.

Weeks later, the deviation was sitting in a review. The record was clean in the way system records are clean, with a date, a timestamp, a location, the expected sequence, the actual sequence, and the measurable difference between the two. The system could show exactly when Marcus had deviated and how long the change lasted. What it could not show was the traffic he had been looking at, the pattern he had learned over eleven years, or the calculation he made in the moment about which choice would get the work done with the least disruption. The record had preserved the event with extraordinary precision while preserving almost nothing about why the event had happened.

Marcus looked at the timestamp and tried to remember the day. There had been hundreds of stops since then, dozens of small adjustments, and more route decisions than he could reconstruct from a line on a screen. He could remember the school crossing because it had frustrated him repeatedly, but he could not remember exactly what he had seen at that particular intersection three weeks earlier. What stood out was not the deviation itself but the fact that something had remembered it after he had forgotten it. The system had kept a version of his day that was more durable than his own memory of living it.

He asked the question people ask when a private action suddenly becomes an official one. Who reported it? Who had thought the change mattered enough to carry forward, and who had decided that this particular stop deserved to appear in a review weeks later? The manager did not have a person's name to give him because no person had reported anything. The monitoring system had seen the deviation, stored it, associated it with his route and his name, and surfaced it when the conditions around the event met whatever threshold made it worth reviewing. Nobody had told on Marcus in the traditional sense, because the system had done something that once required a person to make that choice.

For most of workplace history, getting another person in trouble required someone to decide that what they saw mattered enough to repeat. A coworker, supervisor, dispatcher, or manager had to observe something, decide it was significant, and then carry that information into another conversation knowing that whatever happened next might have started with them. Even when reporting was clearly the right thing to do, there was friction in the act. You had to say the person's name, describe what happened, answer questions, and sometimes face that person later in the hallway, break room, parking lot, or elevator. That social cost did not prevent every report, but it forced a moment of judgment between noticing something and turning it into an organizational event.

That discomfort was not always useful, and it sometimes protected behavior that should have been confronted sooner. People stayed quiet about harassment, safety problems, retaliation, fraud, bullying, and other serious conduct because reporting another person carried personal consequences. Relationships, hierarchy, fear, and loyalty could all distort judgment, which is one reason organizations built systems designed to reduce dependence on whether a person felt comfortable speaking. Removing some of that friction can make workplaces safer and more accountable. The mistake is assuming that all friction was harmful simply because some of it was.

The same friction that sometimes prevented necessary reporting also filtered out an enormous amount of ordinary imperfection. Someone arrived a few minutes late after years of being reliable, and a coworker let it go because one bad morning did not look like a pattern. Someone took longer than expected on a break because something was happening at home, and the person covering the station decided the moment did not need to become a formal issue. A driver changed the sequence because the road in front of him did not resemble the route on the device, and the dispatcher who understood the geography recognized experienced judgment rather than noncompliance. Those decisions were rarely documented, but they were part of how real workplaces functioned.

People were making distinctions constantly between what deserved attention and what deserved escalation. Those distinctions were sometimes inconsistent and sometimes unfair, but they were also forms of judgment built from context. The person looking at the situation could consider who was involved, what had happened before, whether anyone was harmed, whether the action solved a problem, and whether turning the event into a formal process would improve anything. The decision to let something go was not always avoidance. Often, it was the exercise of discretion by someone who understood that not every observable deviation had equal meaning.

A monitoring system does not feel any of that friction. It does not hesitate because it will see Marcus in the break room tomorrow, and it does not wonder whether one out-of-sequence stop is worth damaging a relationship over. It does not know that Marcus covered another driver's route last Thursday when the operation was short two people unless that action exists somewhere in the data and has been assigned value. It does not know that a person who has performed reliably for eleven years may deserve a different interpretation from someone repeatedly ignoring basic expectations unless the system has been explicitly designed to recognize that difference. It sees what it was configured to notice and preserves whatever the configuration tells it is worth preserving.

That is the part of automated monitoring that deserves more attention than the technology itself. Removing the discomfort from reporting does not only remove a barrier to accountability. It can also remove the judgment that once sat between seeing something and deciding whether the thing seen needed to go any further. The system does not have to be cruel for this to matter because cruelty requires intention, and the system does not need any intention at all. The more unsettling possibility is that it is simply indifferent to the distinction people once made between noticing something and deciding that something mattered.

A supervisor walking a floor used to make those distinctions without thinking about them formally. The supervisor might see an experienced worker step away from a station for two minutes and know enough about the person, the shift, and the circumstances to keep walking. Another manager might hear frustration in someone's voice during a difficult morning and understand that one sharp sentence was not the same thing as a pattern of conduct. A dispatcher might see a driver take a stop out of sequence and recognize that the driver knew something about the street the route model did not. Human judgment could be wrong, but it could also see context that did not fit neatly into a category.

The system only possesses the categories someone gave it. If the rule says the station should remain staffed, an absence can become an event regardless of why the person stepped away. If the system has been configured to identify route deviation, the changed sequence can become a signal even when the change improved the actual route. If a monitoring layer is looking for language associated with risk, a phrase in a chat message can become worth surfacing even when everyone involved in the conversation understood what the speaker meant. The system does not have to misunderstand anyone deliberately. It only has to notice continuously and preserve enough of what it notices for the event to acquire meaning later.

That makes this different from the dynamic in "The Dashboard Is Watching." The dashboard changed behavior because people knew their output was visible, and rates, completion times, deviations, response windows, and performance categories created a second audience inside the work. People began thinking not only about what they were doing but about how what they were doing would appear once converted into a number. Automated informing adds another layer because the question is no longer only what the dashboard will show about your performance. The question becomes what the system might notice about your behavior, how long it will remember what it notices, and where that information might travel without another person deciding that it deserved to move.

That movement can extend far beyond productivity. Camera systems can be configured to identify whether safety equipment appears to be missing, whether someone entered a restricted area, or whether a physical movement matches a condition the organization has defined as unsafe. Communication systems can scan language for threats, harassment, confidential information, or other patterns that have been designated for attention. Badge and access systems can record when someone entered a building, which areas they accessed, and how long they remained there. Scheduling systems, location records, performance data, and communication logs can each begin as separate sources of information and later become parts of the same picture.

Nobody has to sit down intending to build a story about one employee for those records to begin forming one. A late badge entry is one fact, a location record is another, a chat message is another, and a performance deviation is another. Individually, none may mean very much, but combined they can look like a pattern even though no human being ever observed the pattern as it was happening. The system can assemble relationships among events that occurred in different places, at different times, and for different reasons. That ability is powerful because it can expose genuine patterns people would otherwise miss, but it can also create apparent coherence where the underlying events were never part of the same story.

That creates a different kind of witness. A traditional witness sees a moment from one position and carries all the imperfections that come with being human. The witness may misunderstand, forget, exaggerate, minimize, sympathize, or decide that what happened was not significant enough to repeat. Those weaknesses are real, but the human witness may also know the person, the circumstances, the history, and the difference between what is normal for that person and what would genuinely be concerning. The automated witness has different strengths and different weaknesses because it may remember everything it was designed to record while understanding very little about what any individual event meant to the people inside it.

Persistence is one of its greatest strengths, and persistence can become authority. Marcus could no longer reconstruct the traffic from three weeks earlier, but the system could reconstruct the timestamp with complete confidence. He could say that he probably had a reason for changing the sequence because he knew how he worked and knew the route, but the record could say exactly when the stop occurred and exactly where it sat against the expected sequence. One version sounded uncertain because human memory is uncertain, while the other looked precise because timestamps are precise. The danger is that precision about when something happened can be mistaken for certainty about why it happened or what the event means.

The imbalance becomes stronger when no accuser exists. Marcus cannot ask what the person who reported him actually saw because there was no person standing there making a judgment about his behavior. He cannot explain the traffic to a witness, ask whether the witness understood the reason for the change, or challenge the witness's interpretation of what happened. He is arguing with a record that contains no human intention and therefore no human perspective that can be revisited. The system did not accuse him in the emotional sense, but it produced the evidence around which an accusation could later be built.

The manager on the other side of the conversation inherits that evidence and a problem the manager did not create. The record exists, the system surfaced it for a reason, and ignoring it may now require more explanation than responding to it. The manager may believe Marcus, understand the route, and know that one small deviation is not particularly meaningful. Even then, the act of surfacing the event has already changed the starting point because something that might once have disappeared inside ordinary judgment now exists as an item that may need to be reviewed, documented, or closed. The system has exercised part of the discretion before the manager enters by deciding what deserves the manager's attention.

This changes more than the relationship between a worker and management. It begins changing relationships among workers because the ordinary cover people give one another becomes less reliable when the system may be observing independently. I do not mean covering serious misconduct or protecting someone who is creating risk for other people. I mean the small allowances that accumulate when people know one another well enough to distinguish a difficult morning from a pattern, or an understandable adjustment from a reckless one. Those allowances are part of the social structure of work even though organizations almost never measure them.

Someone says, I have your station for five minutes, go handle what you need to handle. A manager sees that an otherwise reliable employee is having a difficult morning and gives the person space before turning the moment into a conversation about performance. A coworker notices that someone missed a small step, corrects it quietly, and does not make sure everyone knows who made the mistake. These actions are not automatically favoritism or concealment. Often, they are how people preserve enough flexibility for real human beings to function inside standardized work without converting every imperfect moment into a formal record.

Continuous monitoring changes the value of those gestures because the people involved may no longer be the only witnesses. A coworker can cover the station, but the badge record may still show who moved where and when. A manager can decide that a small lapse does not require escalation, but a monitoring system may already have logged the event and placed it into a pattern. A team can collectively understand why a deviation occurred and still discover later that the system understood only that a deviation occurred. The informal decision to let something go loses power when the formal system has already decided to remember it.

That shift changes behavior before any disciplinary action occurs. Terrence would notice it first in the room rather than in a report because the standup becomes slightly quieter, the questions become more careful, and people start choosing words with the possibility of a second audience in mind.

Nobody announces that trust has changed, and no metric identifies the exact day it happened. People simply become more deliberate about what they say, what they volunteer, and how much of themselves they are willing to expose inside an environment that may preserve a fragment of a moment without preserving the context around it. The caution becomes visible only to someone who remembers what the room sounded like before everyone began calculating how they might be recorded.

Denise would recognize the same shift on the floor. Workers who once solved small problems together begin routing more things through formal steps because informal help can create activity the system may interpret differently later. A person who might once have covered for a coworker starts thinking about whether their own record could become connected to whatever is happening. People hesitate before improvising because an improvisation that makes sense to everyone present can look different when reconstructed from timestamps and categories. The relationship shifts from I know what you mean to I do not know what the system will make of what we are doing.

The system does not have to tell people to distrust one another. It only has to make the consequences of being visible uncertain enough that people begin protecting themselves from interpretations they cannot predict. Self-protection narrows cooperation because people become less willing to step outside the exact boundary of their own task, less willing to improvise on behalf of another person, and less willing to participate in an informal solution that might look different when reconstructed later from data. The work can become cleaner on paper while becoming thinner in practice. Fewer undocumented adjustments may look like stronger compliance even when some of those adjustments were the ways experienced people kept small problems from becoming larger ones.

This is where surveillance begins affecting culture even when the surveillance produces no dramatic event. People do not need to be disciplined constantly for monitoring to change them. They only need to understand that ordinary behavior can be captured, stored, connected, and surfaced later without anyone around them deciding that it deserves to be. The possibility of later interpretation becomes part of the current decision, and people begin managing themselves not only for the people in the room but for the system that may eventually reconstruct the room differently. The organization acquires an audience that is always potentially present and never socially visible.

That audience has no face, which matters because people are good at reading other people. We learn the expectations of a manager, recognize when a coworker is joking, understand which conversations are serious, and adjust to the social texture around us. A system does not offer that texture back because you do not know which minor action will remain minor, which phrase will be interpreted as ordinary language and which one will become a flag, or whether a route adjustment will disappear into thousands of other adjustments or become one point in a pattern assembled later. The uncertainty does not need to be constant or dramatic to shape behavior. It only has to be credible enough that people begin carrying it with them.

People therefore begin managing themselves for the possibility of interpretation. That is different from simply following rules because following a rule means knowing the boundary and deciding to remain inside it. Managing yourself for a monitoring system means knowing that the boundary may later be reconstructed from data you cannot see, using categories you did not define, at a time when the context that made the action reasonable may no longer be available. The safest behavior becomes the behavior most easily explained by the record. Over time, legibility begins replacing judgment in the same way it has elsewhere in the system.

Marcus had already learned this from the scanner. Following the sequence was safer even when the road said otherwise because the sequence was visible and the reason for changing it might not be. Automated informing adds the knowledge that even deviations nobody cared about in the moment can survive until a system decides they belong in a later conversation. The small decision does not necessarily end when the moment ends because the record can keep carrying it forward. That permanence changes the calculation people make before exercising judgment because the immediate situation is no longer the only audience that matters.

The effect reaches discretion at every level. For the worker, discretion means the freedom to solve an ordinary problem without wondering whether the solution will later appear as evidence of noncompliance. For the coworker, discretion means being able to recognize that another person's bad morning does not need to become an official event. For the manager, discretion means being able to see something, understand the circumstances, and decide that the right response is to let it go. Those forms of discretion are different, but they all depend on there being some space between noticing an event and turning the event into a permanent organizational fact.

That last form of discretion may be the one organizations lose without realizing what they have lost. Management has always involved deciding what not to escalate because every imperfect operation generates more mistakes, tensions, delays, workarounds, misunderstandings, and small deviations than any organization could formally process. Good managers filtered what they saw, distinguishing noise from signal, a lapse from a pattern, an understandable adjustment from a reckless one, and a person having one difficult hour from a person creating an ongoing problem. That filtering was part of judgment even though no dashboard measured the number of unnecessary escalations that never happened. An experienced manager deciding that something did not need to go further could prevent the organization from spending time, trust, and attention on an event that did not deserve any of them.

Monitoring systems can weaken that filter without explicitly removing the manager's authority. The manager may still decide what happens after the flag arrives, but the act of surfacing the event has already changed the starting point. Something that might once have disappeared inside ordinary human discretion now exists as an item that someone may need to explain closing. Before continuous monitoring, the question was often whether an event was serious enough to report. After the system reports automatically, the question becomes whether the event is insignificant enough to dismiss, and those are different standards because one requires energy to escalate while the other requires energy to de-escalate.

That distinction becomes more important under volume. A manager facing forty surfaced deviations does not have unlimited time to reconstruct the human context behind each one. The system has already done the easy part by identifying what crossed a threshold, while the harder work of determining what the threshold crossing actually means remains human. Under enough volume, some of that judgment will collapse into process because processing the queue becomes more manageable than reopening the meaning of every event inside it. The flag exists, the response is required, the manager closes the loop, and the distinction between something worth noticing and something worth escalating gets thinner with each cycle.

Marcus had already experienced pieces of this without seeing the full mechanism. He knew the scanner recorded his stops and knew route variance could appear in performance discussions, but he had not fully understood that the absence of a human complaint did not mean the event had disappeared. The system had been present when nobody else was paying attention, and its presence did not depend on anyone caring enough in the moment to remember what happened. It noticed the out-of-sequence stop, recorded the time, connected the event to his route, and preserved it without anger, concern, loyalty, annoyance, or any other human reason for doing so. Weeks later, the record had moved far enough through the process to become something he was expected to explain.

That was the moment the structure became visible to him because the case had not begun when the manager opened the review. It had not begun because a coworker or dispatcher believed Marcus had done something wrong, and it had not begun when someone consciously decided that his judgment needed to be challenged. The first step had happened much earlier, when the system noticed the deviation and preserved it as something capable of becoming evidence later. By the time Marcus saw the record, the event had already survived long enough to acquire institutional weight. What had felt like an ordinary decision on the street had become part of a story nobody had consciously decided to write.

That is why automated monitoring cannot be reduced to privacy in the abstract. Privacy matters, but the operational loss is more specific because what disappears is the space between seeing something and deciding that it needs to become anything more. That space used to contain human judgment, and while that judgment sometimes failed, protected the wrong people, or allowed serious problems to continue, it also allowed ordinary workplaces to absorb ordinary imperfection without converting every small deviation into a permanent record. The challenge is not to restore every form of informal silence. It is to recognize that a workplace without any capacity to let small things remain small is not necessarily more accountable simply because it has more evidence.

A system built to notice continuously has no discomfort to overcome before it remembers. It has no relationship to protect, no elevator ride afterward, and no instinct telling it that a person with eleven years of good judgment may have had a reason for taking one stop out of sequence. If the system is designed to surface the condition, surfacing the condition is evidence that the system performed its function. Whether the condition deserved to become part of someone's record is a different question, and that question still requires judgment even after the system has made the event impossible to forget. The danger is that organizations may confuse the ability to remember everything with the wisdom to know what deserves remembering.

The loss is therefore larger than privacy because the disappearing resource is discretion. It includes the discretion of workers to make small reasonable adjustments, the discretion of coworkers to distinguish a difficult moment from a reportable one, and the discretion of managers to decide that something they noticed does not need to become anything more. Workplaces have always depended on that space more than their policies admit, because human systems cannot function if every imperfection receives the same level of attention. As monitoring becomes continuous, organizations will have to decide whether they intend to preserve some form of that judgment deliberately or allow every observable event to remain available for later interpretation simply because the systems built to notice and remember have no reason, on their own, to let anything go.

Meetings That Don't Decide

By the time the meeting begins, the outcome is already set. The deck was sent the night before. Pre-meetings have happened. Positions have been tested in smaller, safer conversations. Concerns have been softened or traded for something else. What remains is not a question of direction. It is a process of confirming direction in a format that feels structured and complete. The meeting is not where the decision is made. The meeting is where the decision is announced with the appearance of having been made together.

Terrence had not been in one of those rooms before. Not the room at that level. He had run standups, managed shift briefings, sat in the operations reviews where his numbers appeared on slides. But the planning session where targets for his floor were actually built was a different room entirely, and he had not had visibility into it until the week after Margaret asked her question. He had been invited as part of a broader input process, a phrase that, in the organization's language, meant that the decision had already been shaped but that a certain number of floor-level managers would be present when it was presented. He sat down at the table, looked at the deck that had arrived in his inbox the evening before, and understood within the first ten minutes why Margaret had not gotten a real answer.

The metric targets he had communicated to his team on that Tuesday morning had been built months earlier, assembled from productivity benchmarks and historical throughput data and labor optimization models that ran at a level where the specific conditions of his floor had no representation. The logic was clean on the slides. It was the same kind of logic that sounds reasonable when read aloud from a slide, which is exactly how he had delivered it to his team. The numbers were there. The direction was clear. What was not in the room, and had not been in the earlier room either, was the question Margaret had asked: where is the time supposed to come from. The room he was sitting in now had not been designed to ask that question. The room he had stood in front of his team in had not been designed to answer it. The two rooms were connected by a chain of ratification that made both of them, in their different ways, incapable of the conversation that needed to happen somewhere.

I have been in enough of these rooms to recognize the specific quality of air they contain. Nothing feels at risk. Questions are asked, but they tend to clarify rather than challenge. Heads move at expected moments. The person presenting knows exactly where the resistance is and has already addressed it in a pre-meeting two days ago. What the room is doing is not deciding. It is ratifying. And everyone in the room knows it, and almost no one says so, because the function of the meeting is not to think but to produce a record of alignment that can be shared upward.

This pattern exists for legitimate reasons. Large organizations cannot absorb constant disruption. Pre-alignment reduces friction, prevents public conflict, and allows execution to move at the pace required. I am not arguing against preparation. I am arguing against the conflation of preparation with decision-making, against the fiction that a ratification ceremony is the same thing as the room actually working something out.

Real decisions require tension. They involve disagreement, incomplete information, and the willingness to sit with options that do not resolve cleanly. When that work is pushed entirely outside the room, into the calls and messages and corridor conversations where positions are traded before the official conversation begins, something essential is lost. Not the decision itself, necessarily. But the process by which the decision might have been made better: the perspective that did not make it into the pre-alignment, the question that would have slowed things down in a way that would have been worth it.

What Terrence understood in that room was not complicated, though it took him some time to find the words for it. He was not there to provide input. He was there to receive the decision that had already been made, in a format designed to look like input was being gathered, so that the people who made it could document that the floor-level managers had been consulted. He had been consulted. He had not been heard, because there was no mechanism in that room for the thing he knew, which was that the math did not work at the headcount his floor was actually carrying, to reach the part of the process where it would have mattered. That part of the process had already concluded before he arrived.

I remember a specific meeting where something similar happened to me. A decision presented as complete, stakeholders aligned, data supporting the direction. Everything was in place. Halfway through the presentation, I thought of something that had not been part of the earlier conversations. Not a fatal objection. Just a dimension that had not been fully considered. I raised it. The room paused in the particular way rooms pause when someone has introduced uncertainty into a process that had already finished processing. The leader acknowledged the point with the practiced warmth of someone who needed the meeting to continue. The group moved on. The decision went through. I do not know whether what I raised would have changed anything. I know that the structure of that meeting was not designed to find out.

The most consequential conversations in most organizations happen before the meetings that are supposed to contain them. Once people understand that, they redirect their attention accordingly: toward the pre-meetings, the informal access, the early positioning that determines whether a perspective is incorporated into the decision or introduced too late to matter. Influence migrates to where the real work is happening. The official meeting becomes a performance, a necessary one perhaps, but a performance nonetheless. Terrence drove back to his facility from that planning session and stood in his next Tuesday standup with a clearer picture of what had happened the week before. He had not been unable to answer Margaret's question because he lacked the knowledge to answer it. He had been unable to answer it because the answer had been decided in a room he had not been in, by a process he had not been part of, and his standup was the last stop in a chain that had never once asked whether the math worked.

Alignment is what an organization looks like when it has stopped listening. It is the moment the outcome is no longer genuinely uncertain, and therefore the moment the most useful kind of thinking stops being invited.

Alignment Becomes Survival

Direction begins to take shape before it is ever stated outright. It travels through tone, through emphasis, through which metrics get mentioned first in senior conversations and which ones get qualified. A comment from the right person in the right setting carries more information than any formal communication, and the people who need to know, the ones whose careers depend on reading the environment correctly, pick it up immediately. They are not being cynical. They are being professional. They have learned that the cost of misreading the direction is higher than the cost of moving before they have been explicitly asked to.

I have moved with this current, more than once, and for reasons I want to be precise about: not always because I agreed with the direction, and not always because I was afraid. Sometimes because I had done the calculation and concluded that the energy required to resist a direction that was probably going to happen anyway was energy I did not have and could not afford to spend on a battle I was unlikely to win. That calculation is rational. It is also the mechanism by which organizations lose the capacity to hear things they need to hear.

When alignment becomes the default response to perceived direction, when people are adjusting before they are asked, framing ideas to fit within whatever has momentum, filtering their own thinking before it reaches the surface, the organization is thinking only with the intelligence that has already decided to agree. The rest of its intelligence is still present. It has simply learned that being present and being heard are not the same thing.

What fades is not disagreement. Disagreement does not disappear. It relocates. It moves into the private conversations, the messages sent after the meeting, the informal spaces where things can be said without becoming part of the record. I have had extraordinary conversations in parking lots and hotel corridors after meetings that had produced complete apparent agreement. The actual thinking was happening there. The meeting had been the formality. Everyone understood this. Nobody said so.

I was part of those conversations. I have said things in parking lots, hotel corridors, elevators, and after-hours calls that I did not say in the meeting that preceded them, not because the thoughts were less important, but because the formal room had no structure for receiving them. The parking lot did. The conversation after the meeting allowed uncertainty, doubt, frustration, and contradiction to exist without immediately becoming a position someone had to defend. People could say what they actually thought because the conversation no longer carried the same institutional consequences.

I understood, even then, that what was being said outside the room was often more accurate than what had been said inside it. The concerns were sharper. The disagreements were clearer. People named assumptions they had allowed to pass without challenge only an hour earlier. Someone would say that the plan was too aggressive, that the numbers were missing something obvious, or that the team being asked to execute the decision had never really been consulted. The thinking had not disappeared during the meeting. It had simply waited until the environment was safe enough for it to surface.

What mattered was what happened next. Most of the time, nothing did. I was not going to walk back into the room, reopen the decision, and tell everyone that the conversation in the parking lot had been more honest than the one we had just documented. Neither were most of the people standing beside me. We had already made the professional calculation. The meeting had ended, the direction had momentum, and reintroducing uncertainty would require someone to accept responsibility for slowing the process down. The conversation would remain where it had occurred, in the unofficial space where truth could exist without consequence, while the official decision proceeded on the basis of what the room had recorded as agreement.

I did not think of that as dishonesty at the time. I thought of it as professionalism. There were things worth saying and things worth letting go. There were battles that mattered and battles that would only consume credibility without changing the outcome. Experienced people learn to distinguish between them because no one can challenge everything. I believed I was choosing my moments carefully, preserving influence for the places where it would matter most. That explanation felt responsible then. It feels less complete now, because repeated often enough, choosing the moment can become choosing silence.

That is how alignment begins turning into survival. The person is not necessarily agreeing with the direction. They are deciding how much disagreement they can afford to expose. They begin measuring the value of the objection against the likely cost of raising it, the standing of the person they would be challenging, the maturity of the decision, the appetite of the room, and whether anyone with more authority is likely to support them once the discomfort begins. None of those calculations appear in meeting notes, but they shape what the organization is allowed to hear.

The result is that the formal conversation becomes cleaner as the real conversation becomes more fragmented. The meeting shows alignment because the disagreement has migrated elsewhere. It moves into private messages, hallway conversations, calls after the decision, and comments made to people who already agree. The organization can therefore look increasingly unified at exactly the moment its thinking is becoming less integrated. Everyone may know that important objections exist, but knowing privately is not the same as allowing those objections to enter the process where they can affect the outcome.

The cost of that relocation is difficult to measure because the organization never sees most of what it loses. The strongest challenge may never be rejected because it is never formally raised. The operational detail that would have changed the decision may remain in the head of someone who concluded that the room was too far along to hear it. The alternative explanation may be shared with three colleagues who all agree and then disappear because none of them believes carrying it back into the formal process is worth the exposure. The organization proceeds without the missing information and has no way to count what never entered the record.

Over time, people get better at this kind of calibration. They learn which phrases create resistance and which ones sound aligned. They learn to turn disagreement into a question, a question into a clarification, and a clarification into something that can be acknowledged without requiring the direction to change. They become skilled at offering enough friction to demonstrate thoughtfulness without producing enough friction to become identified with the opposition. That skill can look like maturity because sometimes it is maturity. In environments where alignment is heavily rewarded, however, the same skill can become a survival mechanism.

The danger is not that people stop thinking independently. Most of them do not. They continue seeing the flaws, discussing the trade-offs, recognizing what the model missed, and questioning the assumptions among people they trust. The danger is that independent thinking becomes increasingly private while public behavior becomes increasingly aligned. The organization retains the intelligence but loses access to it. People still know what they know. They simply become more selective about where that knowledge is allowed to exist.

That selection process eventually changes the ideas themselves. A challenge that can only survive in the right room, with the right audience, delivered by the right person, at exactly the right moment is not a challenge the organization can reliably use. It becomes dependent on access rather than merit.

People with stronger relationships can surface concerns that others cannot. People with less political capital learn to wait. People closest to the work may understand the problem most clearly while having the least freedom to state it in the form most likely to affect the decision.

This is where useful information begins dying in translation. The original concern may be direct: the staffing plan will not work. By the time it travels upward, it becomes a request to review assumptions. The floor knows the target is unrealistic, but the formal language says there may be an opportunity to revisit the model. A manager believes the plan will create a failure, but the presentation says the team will monitor performance closely during implementation. Each translation makes the concern safer. It also makes it weaker.

Eventually, the organization receives something so softened that it no longer carries enough force to change anything. The person who raised the concern can still say they raised it. The leader receiving it can still say feedback was considered. The process remains complete from an administrative standpoint, while the substance that made the concern worth hearing has been diluted out of it. Alignment is preserved because the challenge has been translated into language that no longer threatens the direction.

There is a version of alignment that every functioning organization needs. Teams cannot reopen every question indefinitely, and leaders cannot execute if every decision remains permanently uncertain. At some point, disagreement has to give way to commitment. People who argued for another direction may still need to execute the one that was chosen, and that is not inherently unhealthy. Mature organizations depend on the ability to debate honestly and then move together once the decision has actually been made.

The problem is not alignment. The problem is premature alignment, when the obligation to support the direction arrives before the organization has finished thinking about what the direction should be. The difference between the two is whether the outcome was genuinely uncertain when people were invited into the conversation. If the answer could still change, alignment can be the result of deliberation. If the answer had already hardened and the conversation existed mainly to gather acknowledgment, then what looks like alignment is something else.

It is ratification with better language. People are invited to contribute after the practical cost of changing direction has become too high. Their presence creates the appearance of participation, their questions create the appearance of scrutiny, and their eventual silence creates the appearance of agreement. The organization then records that agreement and moves forward with greater confidence because more people were technically involved. What the record cannot show is how many of those people had already concluded that meaningful disagreement was no longer professionally useful.

Organizations that mistake that process for genuine decision-making gradually become less capable of correcting themselves. They still hold meetings, collect input, circulate drafts, and ask whether anyone has concerns. The forms of participation remain intact. What changes is whether anyone believes the structure can still absorb a challenge large enough to alter the direction. Once the answer becomes no, people adjust long before leadership realizes they have adjusted.

That is when alignment becomes survival. It is no longer simply the discipline of moving together after disagreement has been resolved. It becomes the discipline of knowing which disagreements should never become visible enough to threaten your position inside the group. The meeting still produces consensus. The organization still produces a record. But the decision represented by that record may have been made somewhere else, by fewer people, before much of the intelligence sitting in the room was ever given a realistic chance to affect it.

The Cost That Doesn't Show Up

The targets were met. The quarter closed well. The report was clean. Standing in that building at the end of a shift that had gone exactly as the metrics said it should, the work felt heavier than it had six months ago. It was not dramatic. It was not a crisis. It was a quality of air, a density in conversations that used to feel lighter, a disappearance of the kind of informal initiative that is never noticed until it is gone. The dashboard said everything was fine. The building told a different story, and the building was right.

Six months earlier, on a Tuesday morning in the middle of a standup, Margaret had asked Terrence where the time was supposed to come from. It was not a challenge. It was a reasonable question from someone who had been on that floor longer than Terrence had been managing it and who wanted to understand how to do the work correctly. The question never got answered. Terrence had restated the expectation in different words and moved on to the number that needed to improve by end of week. The meeting ended. Margaret went back to her station. The question was not revisited then, and it has not been revisited since.

What changed after that is worth describing precisely, because it did not look like deterioration from any angle the system was configured to see. Margaret's pace increased. Her numbers improved. She appeared in the monthly performance summary as a worker trending in the right direction, and she was: her throughput was higher, her error rate was lower, her adherence to the modified pick sequence was consistent in a way it had not been before. What the performance summary did not record was what she had given up to produce those numbers. She had been in the habit, for most of her time on that floor, of noting concerns at the end of her shift on the associate's whiteboard in the break room: recurring problems with specific induct points, package sizing issues that created downstream confusion, the occasional routing error that she caught before it became a logged complaint. She had not added anything to the whiteboard in eleven weeks. Not because the concerns had stopped presenting themselves. Because the whiteboard had stopped generating responses, and raising things that went nowhere had a cost she was no longer willing to pay.

The narrowing was not a decision she made once. It was a series of small ones, each of them rational, that accumulated over the months following the standup into something that looked, from above, like improved performance and felt, from inside, like a different kind of work. She stopped pausing at the station near the east loading dock where the tunnel scanner had a known angle gap on certain package sizes, because pausing cost time and the time had a metric attached to it now and the concern did not. She stopped flagging the weight discrepancies she caught on the high-value sort, because she had flagged them before and the flags had moved through a review cycle that ran slower than the freight and had produced no visible change in the handling instructions. She still caught the discrepancies. She just moved the packages along and let whatever happened next happen downstream. The knowledge she carried about that floor did not disappear. She stopped deploying it. The cost of deploying it had come to exceed the cost of keeping it to herself.

The gap between what a system records and what the work actually costs the people doing it can be significant and sustained before anyone looks for it. That gap is not a malfunction. It is a structural feature of how optimization works: measurable output is the target, and the inputs that generate it, energy, investment, the willingness to go beyond what is technically required, sit outside the frame. The structure performs. The people sustain it. The cost of sustaining it accumulates in ways the structure cannot see and therefore does not address.

Denise had been managing the sort operation for six years. For most of them, her first forty minutes on shift looked the same: she walked the floor, stopping by the breakroom to check the associate whiteboard for anything that had surfaced during previous shifts, not inspecting, not auditing, just moving through the space and watching. She knew which stations ran tight on Tuesdays, which workers went quiet when the line got long, where small problems gathered before they became logged ones. She caught things in those forty minutes that never appeared in any report, because she caught them before they were the kind of thing that appears in reports.

The walk had no code in the shift management system. It registered as unproductive time, idle movement across a floor that her metrics should have been managing from her laptop. When the administrative workload expanded and her data entry compliance scores began showing up in her monthly reviews, she stopped walking, not because anyone told her to, but because the math had shifted and the walk no longer fit inside it.

She still managed the operation, still hit her numbers, still submitted her compliance reports on time. She managed it from her seat now, watching figures that described the floor she used to know by feel, catching nothing until it was already the kind of thing that gets logged.

What she lost is not easy to name. It is the kind of knowledge that does not transfer to a dashboard: that a particular worker's slower pace on a Thursday usually means something is building at home and the next two shifts will need extra attention, that the induct station near the east loading dock door has a gap in its tunnel scanner angle that only shows up on certain package sizes and requires a manual confirmation the system never prompts for. She still knew those things. She just no longer had the mechanism to act on them before they became data.

Three weeks after Denise stopped walking the floor, a picker named Ray caught a processing error on a high-value shipment. He caught it the way experienced workers catch things: something felt off, a weight that did not match the label, a package that had been sorted twice. The old habit was to note it on the associate's whiteboard in the break room, the same whiteboard Margaret had been adding to for years and had stopped adding to eleven weeks before. Six months earlier someone would have responded to a log by end of shift. Ray had watched the whiteboard fill up over the previous weeks without generating a single conversation. So he fixed the error himself, rerouted the package, and moved on. His own small workaround, reasonable under the circumstances, required bypassing the standard verification chain. He did not know, and had no way to know, that two weeks later a similar package on the same route would go out wrong without anyone catching it, because the workaround had become the habit and the habit did not include the check.

The whiteboard kept filling. Nobody read it. It sat there as a record of what the floor knew and the structure did not, accumulating the operational intelligence of people who had learned that raising concerns formally went nowhere and raising them informally went nowhere faster. Ray's name appeared in it six times in a single month. Not because he was struggling. Because he was paying attention to things the system had no category for.

Two months later, a customer filed a complaint. The package had arrived damaged, addressed incorrectly, delivered to a neighbor who had left it on the wrong porch. The customer had been waiting for medication. The complaint moved through the formal resolution system and was logged as a delivery failure attributable to an address verification gap. The root cause identified was human error at the sort stage. The corrective action assigned was a retraining reminder for sort teams on address confirmation protocols. Nobody traced the complaint back to the whiteboard. Nobody traced it back to Denise stopping her walk, or to the administrative load that had shifted her attention from the floor to the laptop, or to the quarterly review that had made the compliance score visible.

Nobody traced it back to Margaret, who had caught three similar routing inconsistencies in the months before she stopped adding things to the whiteboard. The complaint existed as a single data point. The structure had no mechanism for showing that it was the downstream expression of dozens of upstream decisions, each of them rational, each of them made by someone trying to operate correctly inside an environment that had stopped rewarding the behaviors that would have prevented it.

Pressure inside these environments has no single source. It does not arrive through a clear directive or a specific person's demands. It emerges from the structure of expectations, from targets that reset continuously, from baselines that ratchet upward without corresponding changes in conditions, from the fundamental inability of optimization to recognize completion. The next gap is identified before the last one has been closed. The people inside the environment are always moving toward a horizon that keeps moving. The absence of pressure eventually starts to feel like a sign that something has been missed rather than evidence that something has been earned.

What cannot be measured accumulates anyway, but in reverse. The extra check Margaret used to run on the high-value sort. The concerns she wrote on the whiteboard. The pause at the east door station that caught the scanner angle gap before it produced a damaged package. The decision to absorb a problem rather than pass it along. None of those were heroic actions. They were the tissue connecting everything the metrics measure, and when they begin to disappear, gradually, through the accumulated small calculations of people who have learned what the environment rewards, the organization starts losing something it cannot name and therefore cannot protect.

What the performance summary recorded in the six months after that Tuesday standup was improvement. Margaret's throughput was up. Her error rate was down. The dashboard reflected a worker who had responded to the new expectations by scanning more. That reading was accurate as far as it went. What it did not reach was what she had traded to produce it, the specific operational knowledge she had stopped deploying, the concerns she had stopped raising, the check at the east door station that no longer happened. The system recorded the output. It had no column for what the output cost. The cost was real and it was accumulating, in the quality of the work moving through that floor and in Margaret's own relationship to the job, and the only place it appeared was in the building, in that quality of air that was heavier than it had been, and in the question from six months ago that had never been answered and that nobody was going to revisit.

Burnout rarely arrives as a crisis. It arrives as a narrowing. Conversations become more transactional. Initiative retreats to safer ground. The work continues. Outputs remain acceptable, metrics hold. The quality of engagement shifts in ways that are not immediately measurable. The structure records success. The people producing it experience something different. Both are true simultaneously. The one the organization sees is not the one its people are living. Eventually, the metric wins. Not because it is more important. Because it is the only thing anyone is counting.

The Name That Appeared

Ray was not the kind of person who drew attention. He was not the fastest worker on the floor, but he was not slow either. He showed up, absorbed the rhythm of the work, adjusted when conditions changed, and covered the gaps that appeared without waiting to be asked. The operation depended on people like him in the way a building depends on its load-bearing walls: completely, invisibly, and without acknowledgment until something goes wrong. For the better part of two years, nothing went wrong that could be connected to him. He worked inside the acceptable band the system defined and stayed there, which is all the system asked of anyone, and which most people, if they are honest about what the job requires, understand is harder than it sounds.

The first complaint was not about Ray. It was not about anyone. It was a customer report filed through the customer portal, a damaged item photographed and submitted and logged into a dataset that already contained thousands of similar reports, each one processed identically regardless of what had actually happened to the package, who had touched it, or how many hands it had passed through before the damage became visible. The complaint existed in the system as a category, not a story. It had no memory of the conditions that produced it. It had no way to distinguish between a box that was damaged in transit before it arrived at the facility, a box that was damaged by a pallet jack in the receiving bay, and a box that was damaged by a specific person at a specific station during a specific shift. It had timestamps and scan records and location associations, and from those it could construct a sequence. The sequence was not the same thing as the truth of what happened. The system did not know the difference.

What the system could do, and did, was trace the package through the points where it had been recorded. Every scan left a mark. Every association left a record. The package had been in a certain area at a certain time, and in that area at that time, the system's records showed a small number of people. The complaint needed a source, not because anyone had decided to find a culprit, but because the effect of the design was to identify the closest point of association when something went wrong, and associating errors with locations and names was how it generated the accountability data that managers reviewed and that fed into the performance records that determined what happened to people. The system found the closest point. The closest point was Ray.

The second complaint arrived the same way. A different package, a different customer, a different photograph, the same intake process, the same trace through the same dataset. The system found a path through the records and the path led, again, to the same area during the same shift window. Two data points in proximity do not constitute a pattern in any meaningful sense of that word, but they constitute the beginning of one, and the system was built to look for patterns, not to evaluate whether the patterns it found reflected something real or something incidental. It registered the association. The association strengthened the profile. The profile, once it existed, began to accumulate.

Ray knew nothing about this while it was happening. The system was not required to tell him. The data was being processed at a level above the floor, in the dashboards and performance summaries that managers reviewed during the windows between shifts, in the algorithmic scoring that ran continuously and updated the records of every worker in the facility without anyone having to initiate it. What Ray knew was the work. He knew the pace of it, the pressure of it, the physical reality of moving through a facility at the speed the system required, which meant that the careful double-check he had once done on every package at a certain station had become a single check, and then a check he performed when the pace allowed, and then something he did on the packages that looked concerning. He had not been told to stop. The speed had simply made the stopping too expensive to sustain, and the system that created the speed was the same system that was now interpreting the consequences of it as a pattern associated with his name.

The third complaint went through the same process. The fourth. At some point, and there is no exact threshold because the system does not announce when it crosses one, the accumulation became a signal. Not a possibility, not a question worth investigating, but a signal of the kind the system surfaces to indicate that something needs to be corrected. The number of complaints associated with Ray's profile had moved from within the range of what was considered normal to outside it. That movement was what the system was built to detect. It detected it. It surfaced it. And it presented the information to the people responsible for acting on it in the only form it had available, which was a set of numbers attached to a name, with no account of how the numbers had been generated, what they were missing, or whether the name they were attached to bore the relationship to the underlying problem that the numbers implied.

The conversation that followed was not a conversation in the sense that word usually means. It was a review. The data was on the screen. The pattern was visible. The manager who conducted it was not a person looking for someone to blame. She was a person operating inside the same system that had produced the output she was looking at, working from the same training that told her a signal of this was kind as something requiring a documented response, understanding that her own performance included metrics about how quickly and consistently she addressed flags of this type. She presented Ray with what the system had found. He tried to explain what the system could not see. He knew, because he had been doing the work long enough to know, that damage to packages does not always originate where it is discovered. That compromised items arrive from upstream, from carriers, from handling that happened before the facility was involved. That the pace of work the system itself required made the kind of careful inspection that would catch early damage impossible to perform at the rate it needed to be performed. That he had flagged this, on the associate's whiteboard, in the notes field on his scanner, and that none of those flags had generated a response that reached him.

By the time anyone spoke to him directly, the decision had already hardened into something that felt official. Not correct. Not confirmed. Just irreversible. He was not being asked what happened. He was being informed of what the system had already decided happened. The structure left little room for context to change what the record had already established, and the review began from the system's conclusion rather than from an open question about what had happened.

His explanation was consistent with what experienced floor managers often know about those conditions. She had been on the floor long enough to know that the system's picture of what caused damage was incomplete in exactly the ways Ray was describing. But she was not evaluating his account against the system's account to determine which was more reliable. She was managing a documented signal that required a documented response, and his explanation, however accurate, existed in a form the system could not receive. It was not in the data. It could not be entered into the data in any form that would modify the pattern the system had already established. What the formal process gave her was the record, and the record said what it said. Ray's explanation could help her understand what had happened, but it did not carry the same weight inside the system she was expected to act through. She wrote him up. The language was formal and procedural and matched the format the system required. The gap between what Ray knew and what could be documented remained exactly where it was, unchanged, invisible to anything that would matter.

He kept working. The monitoring agent kept working too, continuously reading scan records, complaint data, badge associations, and station assignments for patterns that crossed a response threshold. During the weeks that followed, both Ray and the system were accumulating a record of what was happening, but only one of them had visibility into the other. Ray knew the conditions on the floor, the pressure behind the decisions he was making, and the small adjustments required to keep the work moving. The system knew only what those adjustments looked like once they became data, and it continued building a pattern from fragments Ray did not know were being connected. By the time that pattern surfaced again, the system would have weeks of evidence assembled around his name, while Ray would be left trying to reconstruct the circumstances behind moments he had already moved past.

The Manager Without Power

The outward markers remain intact. Title on the door, team to lead, outcomes to deliver, seat in the room. From a distance, the authority looks real. From inside it, the gap between the appearance of authority and its actual substance is one of the defining experiences of managing inside a system-driven organization, and one of the least acknowledged.

Terrence had been managing his floor for six years by the time he sat across from me and said it plainly: I am responsible for an outcome I am not allowed to fix. He had the job title. He had the team. He had the weekly reviews, the performance targets, and the escalation pathway the organization provided for raising concerns. What he did not have was the ability to change the conditions that were producing the outcome the system was now holding him accountable for.

The metric expectation he had communicated to Margaret and the rest of his team, the one that did not add up against his actual headcount, had been built from benchmarks that did not account for the specific composition of his floor. He knew this because he lived inside the difference every day. He knew which stations ran tight, which workers were still developing speed, which parts of the process required informal coverage, and which assumptions looked reasonable only from far enough away.

He had raised the issue through the appropriate channels. Twice. The response he received was not disagreement. Nobody told him the math on his floor was wrong. Nobody argued that his concern lacked merit. The answer was procedural: changes to standardized processes required a review cycle that ran quarterly, and the next cycle was ten weeks away. He would need to manage within the existing framework in the meantime.

That phrase, in the meantime, contained almost everything. The work did not pause for the review cycle. Margaret still had to process the volume in front of her. Terrence still had to stand in the Tuesday standup and explain the expectation. The dashboard still compared the floor against a target built from assumptions the floor could not meet cleanly. The system had received his concern, placed it into a timeline designed for consistency, and left him to absorb the gap until the timeline caught up.

I did not have a useful answer for him. What he had identified was not a communication failure or a problem of insufficient escalation. It was the structural condition of his role: accountability concentrated at his level, authority to change conditions residing somewhere else, and a formal process for closing that gap calibrated to the needs of the system rather than the needs of the moment.

He had done everything the organization would later say a responsible manager should do. He had identified the issue, documented it, submitted it through the right channel, and continued to manage the operation while the concern moved through the process. The problem was that the process treated the concern as information for a future adjustment, while the floor experienced it as a current condition. Terrence was standing in the space between those two clocks.

That space is where the modern manager lives more often than most organizations admit. The manager is close enough to see what the system is missing, but not always senior enough to change the system that is missing it. He is responsible for translating targets into action, but not for designing the targets. He is accountable for performance, but not always empowered to adjust the conditions performance depends on. The role carries responsibility downward and explanation upward, while the authority needed to resolve the contradiction often sits outside his reach.

Power, in that context, does not disappear. It becomes conditional. It shows up in how effectively a manager can navigate the system, translate its language, avoid unnecessary friction, and produce acceptable outcomes inside constraints he did not create. The managers who thrive are often the ones who learn to work fluently inside those constraints. The managers who struggle are not always less capable. Sometimes they are the ones still trying to do the job as if the title carried the authority the organization's language implies.

Terrence was trying to manage that difference honestly. He had been to the planning session. He had seen where the targets came from. He understood the structure well enough now to know that what he was pushing against was not a single policy that could be revised in the moment, but a design that had already absorbed the possibility of his objection and assigned it to a review cycle. The system had a place for his concern. It did not have a place for his concern to matter soon enough.

Over time, the role shifts in subtle but significant ways. The job becomes less about directing and more about translating: carrying the system's requirements to the team and carrying the team's reality back to the system. In the space between those two functions, the manager absorbs the permanent gap neither side fully acknowledges.

That is what Terrence was doing every Tuesday morning. He was standing in front of Margaret and the rest of his team, communicating expectations he had not designed and could not adjust. He was submitting concerns through channels that responded with timelines. He was the person absorbing what neither side would close. The system had a name for that position. It was manager.

Downward Accountability

Strong performance validates the process. This is one of the most durable and least examined principles in how organizations think about accountability. When numbers are good, the structure that produced them is confirmed. The people who executed within the structure may be recognized, but the system itself also claims the result. Nobody asks too carefully whether the outcome was produced by the structure or in spite of it, because success makes the distinction easy to ignore.

When performance slips, accountability moves in only one direction. The metric flags a gap. The gap has a location. The location has a manager. The manager is now responsible for correction, and the first question is usually what happened at the point of execution, not whether the conditions created upstream made the gap likely.

That movement is not accidental. Large organizations need clear lines of accountability near the work because performance variance has to be addressed quickly. Someone has to own the response. Someone has to explain the miss, correct the behavior, close the loop, and restore alignment. Without that mechanism, complex operations would drift. The problem is that the same mechanism that makes execution accountable also makes design failures look like execution failures.

A staffing assumption becomes a missed rate. A forecast gap becomes a floor-level productivity issue. A routing decision becomes a driver variance. A process design flaw becomes a coaching conversation. By the time the system identifies the gap, the visible failure is usually closest to the person executing the work, so that is where the pressure lands first.

The system is much better at locating variance than locating origin. It can identify where a number went red faster than it can identify the chain of decisions that made the red number predictable. That difference matters because the person closest to the red number becomes responsible for fixing what may have been built several layers above them. The system asks for correction at the point where the failure appeared, even when the cause sits somewhere else.

I have watched capable managers take ownership of outcomes they knew were shaped by conditions beyond their control. They did it not because they were confused about the cause, but because they understood the cost of explaining it too directly. To name the structural cause too early, too clearly, or without the kind of proof the system recognizes can make a manager sound defensive, misaligned, or unwilling to own the result. Better to absorb the pressure and be seen managing through it than to surface the design flaw and be seen making excuses.

That is the trap. The organization tells managers to own the outcome, but not every outcome is fully theirs to own. It asks for accountability at the level of execution while protecting the design choices that created the conditions for execution. The person who can see the gap most clearly often has the most to lose by naming it plainly, especially when the system has already attached the gap to their area, their team, or their name.

Responsibility flows downward because that is where execution lives. Authority to redefine the conditions stays higher up, moves more slowly, and is protected from the urgency that drives the accountability conversation below it. The closer a person is to the work, the faster the system expects a response. The farther a decision is from the work, the more time and insulation it receives before being questioned.

This is how the system sustains itself without requiring bad faith from any individual person. The manager absorbs the miss. The corrective action is documented. The process remains intact. The conditions that produced the miss may be reviewed later, but the immediate accountability has already landed. By the time the organization is ready to examine the design, the person closest to execution has already paid the first cost.

Downward accountability is efficient. It is also incomplete. It keeps the work moving, but it often keeps the wrong questions from moving upward with enough force to matter. The system assigns responsibility quickly because speed is what it values. Producing real accountability would require something slower and more uncomfortable: asking whether the failure belongs only to the person nearest the gap, or also to the structure that made the gap inevitable.

When Integrity Becomes Risk

The situations are rarely dramatic. A process specifies one thing; the circumstances in front of you require another. The gap is small enough that most people would not notice it from outside. But the person inside it notices it precisely, because they are the one who has to decide: in real time, without consultation, with full awareness of how each choice will register, whether to serve the system's definition of the correct action or the situation's actual need.

I was managing a situation once where the technically correct process, applied as written, would have produced a customer outcome that I knew was wrong. Not catastrophically wrong. Wrong in the particular way that erodes trust quietly, the way that produces a customer who does not complain but also does not come back. The deviation required to produce the right outcome was small. It was also outside my documented authority. I made the call, handled the situation as the situation required, and then spent an uncomfortable amount of time considering how to document it in a way that was accurate without being an invitation to a conversation I did not want to have.

That documentation exercise taught me something. The discomfort was not about whether I had done the right thing. I was confident about that. The discomfort was about the gap between what I had done and what the system would recognize as having been done correctly, and the energy I was now spending to manage that gap, to translate a good decision into the language of compliance. That energy is not free. Multiplied across thousands of managers making similar calculations daily, it is a significant tax on judgment, paid in the currency of cognitive load and quiet demoralization.

Integrity, in an environment like this, does not disappear. It gets calibrated. People learn where it can be expressed without consequence and where it carries a cost. They develop a working map of which deviations will be absorbed and which will be noted, which contexts allow judgment and which require strict adherence. That map is never written down. It is learned through observation, through the accumulated evidence of what happens to people who acted on their understanding of what was right at the wrong moment in the wrong context.

The result is not an absence of integrity. It is integrity operating under conditions of managed risk, expressed selectively, in the moments where it is safe, in the language the system already recognizes. That is a different thing from operating with full integrity. The distance between those two things is not always large. It is always real. And in organizations where the system is sufficiently dominant, that distance can become the defining condition of what it means to try to do the right thing inside a structure that is not always designed to make that easy.

The system does not ask people to abandon their values. It asks them to be strategic about when they use them. Over time, strategic integrity and compromised integrity can start to look like the same thing.

I know this because I have sat with the documentation from that situation and been genuinely uncertain which discipline I had practiced more carefully. Three things were true at once. The decision was right, the customer outcome was better than what the process would have produced, and the record was accurate. But it was accurate in the careful way records become accurate when someone understands not only what happened, but also what the system is capable of recognizing and what it is likely to punish. I had told the truth, but I had also translated that truth into a form designed to survive the structure around it.

What stayed with me was how much of my attention went into that translation. The difficult part should have been making the judgment in the moment, weighing the customer's need against the written process and deciding which obligation mattered more. Instead, the harder work came afterward, when I had to think about how the decision would look once the context had been stripped away and the record stood on its own. I reviewed the wording, considered what might trigger questions, and made sure the documentation explained enough without inviting an interpretation that would turn a sound decision into a procedural problem. None of that changed what had happened, but it changed how much energy I spent protecting the decision after it was made.

That is what these environments do to people who are trying to operate with integrity inside them. They do not usually ask you to abandon your values outright, because that would be too obvious and too easy to resist. They teach you instead to calculate the cost of acting on those values, to recognize which moments are safe for judgment and which ones require the protection of perfect documentation. Over time, integrity becomes something you do while simultaneously managing the risk created by doing it. The danger is not that people suddenly stop caring about what is right. The danger is that so much attention becomes devoted to making the right decision defensible that defending it begins to feel as important as the decision itself.

I had managed that gap successfully. I had done what I believed the situation required, produced a better outcome, and created a record the system could accept. What unsettled me was how satisfied I felt with the precision of that management, because it meant I had become skilled at something I should never have needed to become skilled at in the first place. I was learning how to preserve integrity without exposing myself unnecessarily to the consequences of exercising it. After enough repetitions, that skill can begin to look like professionalism, even when what it really reflects is an environment where doing the right thing and safely documenting the right thing have become two separate jobs.

The distinction matters because every time those two jobs separate, some amount of judgment is diverted away from the work itself and toward the system that will later evaluate the work. People begin thinking not only about what should be done, but about how the action will appear when reduced to a field, a category, a timestamp, or a review. That additional calculation rarely announces itself as compromise. It feels responsible, prudent, even necessary. But when the safest version of integrity is the version that has already been shaped around what the system will tolerate, the risk is no longer simply that a good decision might be questioned. The deeper risk is that people eventually stop making decisions they cannot easily defend, even when those are the decisions the situation most needs.

The Copy-Paste Economy

Something worked in one place, so the organization did what organizations do. It extracted the process, documented it, and shipped it everywhere. The logic was sound. Why solve a problem twice when the first solution was sitting there, validated and ready to deploy? The efficiency of replication is one of the genuine advantages of systematic thinking. The problem is the assumption embedded inside it, that what worked in one context will work in another because the contexts look similar from a sufficient distance.

I watched a process get deployed across six locations that had been assessed as comparable. Two performed as expected. Two showed modest improvement. Two created friction that did not show up immediately in the metrics but was visible to anyone who spent time in those buildings. The process was not wrong. It was calibrated for conditions that were slightly different from the conditions it encountered. Different team composition. Different informal hierarchies. Different local knowledge that had built up around the previous process and was not captured in any documentation. The system treated the variation as an execution problem. The people on the ground knew it was a fit problem. The system could not act on the latter, so it acted on the former, which meant applying more pressure to the people carrying the most friction. That made the friction worse.

Processes travel well. Context does not. What gets shipped is not the territory. It is the map: the visible structure of the process, the steps, the handoffs, the ownership model, the documented standard. What rarely survives translation is the invisible infrastructure beneath a successful process, the relationships, the informal knowledge, the local adaptations that were never written down. What survives is the structure. What gets lost is the reasoning that made the structure work.

The system reinforces this loss by treating variation as deviation. When results differ from what the template predicted, the question is almost always what went wrong in implementation rather than what the template missed about the context. That framing protects the template. It also prevents the template from getting better, because feedback that would improve it gets categorized as failure rather than information. The organization learns, very efficiently, how to identify implementation problems. It becomes progressively less capable of learning from them.

Extended faster than adapted became the working motto of the copy-paste economy. Organizations move quickly, maintain consistency, and scale what works. What they lose, incrementally and invisibly, is the sensitivity to difference that would allow them to recognize when what worked before has stopped working now. The map keeps getting shipped. The territory keeps changing. Nobody is formally responsible for the gap between them, which means the gap persists until it is large enough that the system can no longer explain it as an execution problem. By then, several versions of the map have already been sent to the next six locations.

I was the person who signed off on the comparability assessment for those six locations. The assessment was real, not invented, and the dimensions it measured were the right dimensions to measure. What I did not do, and what I understood at the time I was not doing, was spend meaningful time in the buildings I was certifying as equivalent. I reviewed the data. I looked at the operational profiles. I knew that two of the locations had different team compositions from the control site, and I told myself the process was robust enough to absorb that variation. It was not. What I had done was make the assessment the process required and move on. What made it feel like due diligence rather than the shortcut it was, was that I was doing it exactly the way the process had been designed to be done. The comparability framework did not require me to walk those floors. So, I did not walk them.

Algorithmic Management Everywhere

The shift is already complete in many organizations, though it is rarely described in terms that make it visible. Management, the actual work of shaping behavior, directing attention, defining what success looks like and what the consequences of falling short are, has migrated into systems that operate continuously, at a scale and speed that no human manager can match. The algorithms do not hold one-on-ones. They do not give feedback in the traditional sense. They do something more fundamental: they configure the environment in which people work, defining what is visible, what is rewarded, and what triggers a response. Everything downstream of that configuration is, in a practical sense, managed.

I was in a conversation once where a recommendation generated by a proprietary system entered the room with more authority than any individual interpretation of the conditions. We understood what the model was designed to accomplish and knew the broad categories of information behind its output, but none of us could fully reconstruct the reasoning that had produced the recommendation in front of us. I accepted that structure along with everyone else because the organization had already decided the system was reliable enough to use. What stayed with me was the realization that management had begun to mean something different. We were no longer only interpreting information and making decisions from it. Increasingly, we were receiving decisions that had already been partially formed somewhere inside a process we could use more easily than we could explain.

The opacity is the underacknowledged problem. Traditional management, whatever its flaws, was legible. You could identify the person who had made a decision, ask them why, and evaluate the reasoning. Algorithmic management produces decisions without visible reasoning, or rather, with reasoning so embedded in the model's architecture that it is not accessible to the people the decision affects. This creates a new kind of accountability void: outcomes that clearly had causes, generated by a process that resists examination.

The algorithm is not neutral. It was built by people who made choices about what to optimize for, what data to include, what patterns to reward. Those choices are values statements. They reflect a set of priorities that were current at the moment the model was designed and that may not be current now. The model does not update its values when the organization's values change. It produces outputs consistent with what it was built to produce, and if those outputs are close enough to what is expected, the underlying values embedded in the model go unexamined indefinitely.

This is not a reason to reject algorithmic management. The scale at which modern organizations operate makes some degree of algorithmic coordination not just efficient but necessary. The argument is not against the tool. It is against the uncritical adoption of the tool's apparent objectivity, against the habit of treating an algorithm's output as if it were a fact about the world rather than a reflection of the choices embedded in the model. Those choices deserve the same scrutiny as any human decision. They receive, in most organizations, considerably less.

You are no longer disagreeing with a person. You are disagreeing with a system. The system is confident about things it does not understand. That confidence is enough, not because it is earned in this moment, but because it was earned somewhere else and carried forward without question.

When the system is everywhere, accountability is nowhere. And when accountability is nowhere, the question of who is responsible for outcomes produced by systems that nobody fully understands becomes one of the defining challenges of modern organizational life.

The shift from algorithmic management to prediction-based management did not feel like a change in kind. It felt like an upgrade. The tools were simply better, the outputs more sophisticated, the recommendations arriving with greater precision and backed by more data than any previous system had deployed. What changed was not the relationship between the organization and its tools. What changed was the direction of authority. Algorithmic management had looked backward: here is what happened, here is where it fell short, here is the pressure that follows. Predictive management looked forward: here is what will happen, here is what the situation requires before reality has confirmed it, here is the action the model recommends before the moment has arrived to test it. The organization stopped being managed against what it had done. It started being managed against what a system believed it was about to do. The decision was already made before anyone walked into the room. You were being asked to execute inside it.

This did not remain a story about one kind of organization or one kind of industry. The pattern described in these pages, the migration of authority from people into systems, the gradual replacement of human judgment with algorithmic confidence, the structural conditions in which raising what you know is more expensive than following what the model says, appears to be moving. Clinicians describe algorithmic triage and capacity models that shape care decisions they were not asked to design. People applying for loans describe systems whose variables they cannot see and cannot challenge. People navigating schools and benefit programs and housing applications describe the same confident opacity that the logistics systems described in this book deployed years earlier, and often with considerably less accountability when the model turns out to be wrong. The people inside those systems are describing the same experience this book has been examining. That is not proof the mechanisms are identical. It is evidence the pattern is not confined to one industry, and it is worth taking seriously on those terms.

The pressure that built inside companies like this one was not unique to any single company. It was a leading edge, a place where the scale required these tools before most industries understood what the tools would do to the people inside them. Understanding what that transfer cost inside one kind of organization is preparation for understanding what it is costing, right now, in every organization that has decided the efficiency gains are worth whatever the fine print says. The fine print is becoming visible. The cost is becoming general. The system this book has been examining inside one kind of building is now appearing across many of them.

The Black Box Was Under Contract

The recommendation was already on the screen when the conversation began. It had been generated by a proprietary model, supported by a set of inputs extensive enough that nobody in the room had reviewed all of them, and summarized in documentation that gave us enough information to use the output without giving us enough information to reconstruct exactly how the system had reached it. The recommendation looked reasonable, and it had performed well enough in comparable situations that nobody entered the room expecting to challenge the system itself.

I had reviewed the same summary documentation everyone else had. I understood what the model was generally designed to do, knew the categories of information it relied on, and knew enough about the operation to follow the logic of the recommendation. What I did not know was whether the historical comparables behind it included facilities with the same induct configuration as the one we were discussing, because that configuration had changed six months earlier and I was not certain the change was reflected in what the model considered normal. I had a question, and I did not ask it. The room already had a direction, the model had already produced a recommendation, and slowing the conversation down to examine an assumption that might or might not matter carried a cost I understood. I told myself I would raise it afterward, which was one of those promises that feels responsible in the moment because it allows the meeting to continue without forcing you to admit how unlikely the follow up may be.

There was no follow up, and nothing happened that forced one. The recommendation was implemented, the results remained within acceptable ranges, and the organization treated the sequence as evidence that the system had worked. Because the outcome did not create a visible failure, nobody had a reason strong enough to return to the question of whether the reasoning had been sound. For years, I thought about situations like that primarily as failures of human judgment. I should have asked the question. Someone else in the room could have asked it, and the people responsible for the decision could have required more explanation before allowing the recommendation to become action. All of those things remain true, but they leave out something important about the structure surrounding the decision.

We were making a consequential decision from a system whose reasoning was not fully available to the people using it. That limitation was not necessarily something anyone in the room could solve by simply becoming more curious, more skeptical, or more technically sophisticated. The opacity was not only inside the model. Some of it existed in the relationship between the organization and the technology it had decided to trust.

A black box is usually described as a system whose internal reasoning cannot be easily interpreted. Information goes in, an output comes back, and the people using the result understand less about what happened between those points than they would if a person had made the same judgment and explained the reasoning. Inside an organization, that description becomes more complicated when the people relying on the output do not fully own or control the system producing it. Companies rarely build every piece of technology they use. Capabilities are purchased, licensed, configured, integrated, and connected to internal systems because building everything internally would be impractical and unnecessary. There is nothing inherently wrong with that arrangement, and much of modern business could not function without it.

The important question begins later, after the purchased capability starts carrying authority. A company can license software and still treat it as a tool, one source of information among several that a person considers before deciding what should happen. The relationship changes when the system's output becomes the starting point for decisions that managers, workers, and leaders are later expected to defend. At that point, the organization has imported more than technology. It has imported assumptions, classifications, thresholds, definitions, and methods of interpretation that may not be fully visible to the people experiencing their consequences. The system can become part of how the organization decides what is normal, what is unusual, what requires attention, and what counts as evidence before the people closest to the work have had any meaningful opportunity to examine those choices.

The contract may define what the system provides. It may define service levels, support obligations, data handling, implementation responsibilities, availability, security requirements, and the boundaries of what the provider is expected to disclose. Those provisions can be detailed and still fail to answer the question a manager eventually needs answered when an output does not match the operation in front of them. The manager's question is usually much simpler: Why did the system decide this, and can we stop acting on it until we understand whether it is right? Those questions sound operational because they are. Their urgency bears little relationship to the legal or technical structure surrounding the system.

Consider a staffing recommendation that arrives below what an experienced manager believes the floor requires. The manager can see the workforce in front of him, the newer employees still developing speed, the unresolved equipment problem, the shift composition, and the informal adjustments experienced people are already making to keep the operation moving. The system sees whatever variables it has been configured to see and produces the number it was designed to produce. The manager asks why the recommendation is lower than what the floor appears to require. If the organization designed and controls the entire system internally, there may be a difficult but identifiable path toward an answer. Someone owns the model, someone can inspect the assumptions, someone can determine which version is running, and someone can ask whether current conditions still resemble the conditions on which the recommendation depends.

That path may be slow. It may require escalation across teams that speak different technical and operational languages, and the people who built the original model may no longer be there. Even so, the knowledge required to investigate the question exists somewhere inside the same organizational boundary. A proprietary external system can add another boundary. The floor manager asks an analyst, the analyst asks an internal technology team, the technology team reviews what it can see, and the question eventually reaches a point where a complete answer may depend on information the organization does not possess directly. What started as an operational question can become a support request before anyone has actually answered the operational question.

That transition matters because support processes and operational consequences run on different clocks. The worker is standing on the floor now. The staffing decision has already shaped the shift, the route has already populated, or the recommendation has already entered a review. The people responsible for execution still have to act while the question about why the system behaved as it did moves through another process. The system can therefore act much faster than the organization can explain it. That imbalance becomes more consequential as systems move from describing work to shaping it, because every hour spent seeking an explanation may be an hour in which the output is already changing people's schedules, workloads, priorities, or performance records.

This problem is not unique to outside technology. Large internal systems can become just as difficult to understand after years of modifications, integrations, ownership changes, and accumulated assumptions. The difference with a contracted capability is that some portion of the knowledge, access, or authority required to interrogate the system may exist beyond the people who will ultimately be held responsible for what the system produces. The organization controls the decision to use the system, but it may not control every layer of the system it has chosen to trust. Those are not the same thing. That distinction becomes easy to ignore while the technology is producing acceptable results because nobody has a reason to test how much understanding or control the organization actually retained.

The problem becomes visible when something needs to be explained. If a manager makes a judgment call, someone can ask what information was considered and why the choice was made. If an internal team creates a benchmark, someone can identify the group responsible for the methodology and ask whether the assumptions still match current conditions. When a proprietary system enters the chain, the explanation can become distributed across organizations. The provider can explain what the product is designed to do, the internal technology team can explain how it was configured, operations can explain how the output was used, and leadership can explain why the organization decided to deploy the capability in the first place.

Every one of those explanations can be accurate and still fail to answer the question that matters when a real person is standing at the end of the decision asking why the outcome happened. Responsibility has been divided into pieces even though the consequence arrived whole. The provider may conclude that the model operated within its design. The company may conclude that the system was implemented according to the agreed configuration, while the internal team confirms that the data feed remained healthy and the manager confirms that the recommendation was followed correctly. The entire chain can function exactly as intended and still produce an outcome that is wrong for the situation in front of it.

That possibility is difficult for organizations because technical correctness is easier to establish than operational correctness. A system can be functioning as designed while the design itself no longer fits the conditions in which it is operating. When that happens, the ordinary mechanisms for identifying failure may never activate because nothing has failed according to the definitions built into the system. This is one of the ways contractual boundaries can reinforce technical opacity. The problem does not require a provider hiding information or an organization failing to negotiate aggressively enough. There are legitimate reasons companies protect proprietary methods, and there are practical limits to how much technical detail every customer can reasonably inspect.

The problem begins when access to a capability is mistaken for ownership of the reasoning behind the capability. The organization may own every consequence of the decision while possessing less access to the mechanism that helped shape the decision than the people affected by the outcome would reasonably assume. That arrangement matters most when the system begins influencing decisions about people. A proprietary model used to estimate an internal trend carries one kind of risk. A system that influences staffing, work allocation, performance flags, routing, risk classifications, or the actions managers are expected to take carries another because its output is no longer remaining inside an analytical exercise.

Once an output changes someone's day, opacity becomes operational. The question is no longer whether the model is technically interesting or sufficiently sophisticated. The question is whether the organization can explain and challenge the decision quickly enough to prevent a bad output from becoming somebody else's consequence. The system does not need formal authority for this to happen. Institutional confidence can supply that authority long before anyone writes a policy saying the output should be treated as decisive. The fact that the organization selected, approved, purchased, tested, integrated, and deployed the system gives every future recommendation a history the individual manager does not possess.

Someone evaluated the capability. Someone approved the investment, someone connected the technology to company systems, and someone decided the output was trustworthy enough to enter the workflow. By the time the recommendation reaches the manager, the technology has already survived questions the manager was never present to hear. That history travels with the output. The manager is not simply disagreeing with a number on a screen. The manager is disagreeing with a number produced by something the organization has already decided is credible enough to purchase, integrate, and place inside an operational process.

The system therefore enters the room carrying institutional sponsorship. The manager enters carrying what happened on Tuesday, what the floor looks like now, and what experienced people believe is about to go wrong. Those forms of evidence do not begin with equal authority even when the manager is right. I have already described what organizations tend to do with that difference. They ask whether the floor level concern is reflected in the data, whether there is enough evidence to justify a deviation, and whether the local condition is significant enough to override a standardized recommendation. Those questions are reasonable, but the proprietary black box adds another complication because the person being asked to produce evidence may be challenging a system whose complete reasoning the organization itself cannot readily reconstruct.

The manager can therefore find himself being asked to prove the system wrong without being able to inspect everything the system used to decide it was right. That is not a balanced contest, even if nobody involved intends it to be unfair. The imbalance becomes more durable when the system produces acceptable results. In the meeting I described, the recommendation was implemented and the outcome remained inside the expected range. That ended the inquiry because acceptable performance is one of the strongest protections any system can receive. Nobody needed to understand the reasoning after the result looked good enough. The absence of a visible failure became evidence that the system deserved continued trust, even though the original question about the model's comparables had never been answered.

This is how opacity survives ordinary success. The system produces an acceptable answer, the organization acts on it, and the result does not create enough pain to justify the time and disruption required to investigate how the answer was produced. Each apparently successful cycle reduces the incentive to understand the reasoning behind the previous one. Eventually, the organization learns to trust the output instead of the process behind it. That can function for a long time because many systems are useful precisely because they process complexity that no individual person could reproduce manually. Requiring every manager to understand every internal mechanism would remove much of the value that sophisticated technology provides.

The issue is not whether every person should understand everything. The issue is whether somebody with meaningful authority can understand enough when the output stops matching the world. There has to be a point in the structure where operational reality can force the system itself to become the subject of investigation. Without that point, every mismatch becomes somebody else's problem. The provider examines whether the platform operated according to specification, the technology team examines whether the integration failed, operations examines whether the process was followed, and the manager examines whether the worker executed correctly. Each layer looks at the part it owns while the possibility that the reasoning itself may be wrong moves between those layers without finding a permanent owner, particularly when the system has not failed in a way any individual team is responsible for detecting.

That is the accountability problem the contract cannot solve by itself. Contracts can allocate responsibilities between organizations, but organizational responsibility is not the same thing as responsibility for a particular outcome. A document can define who maintains the system, who provides data, who responds to incidents, and who owns various categories of risk without answering who should stop the decision when the system is confidently wrong in a way nobody anticipated. The people downstream do not experience that ambiguity as governance. They experience it as work, because Marcus does not care which technical team owns the routing logic when the sequence places him at the school crossing at the wrong time. Margaret does not care whether the productivity benchmark originated internally, externally, or through some combination of systems when the rate expectation does not fit the conditions on her floor.

Terrence does not need an architecture diagram when the staffing number is wrong. He needs to know whether the number can be changed before the shift that will absorb the consequences. That is where the distance between contractual authority and operational reality becomes visible. The person closest to the consequence usually has the least useful access to the system that produced it. Marcus can see that the route is wrong but cannot inspect the logic that produced it. Margaret can see that the rate does not fit the work but cannot examine every assumption beneath the benchmark. Terrence can see that the staffing plan will create a gap, but the ability to investigate the system may exist several organizational levels away from the place where the gap will be experienced.

If the underlying capability is proprietary, another layer may exist beyond even those levels. The people who understand the operation are separated from the people who understand the technology, and the people who understand the technology may still be separated from the people with access to the deepest parts of the system. Information has to cross those boundaries through documents, support processes, technical interfaces, contractual relationships, and translations between people who describe the same problem differently. Each translation creates the possibility that some part of the original operational meaning will be reduced before it reaches the person capable of changing anything.

Where is the time supposed to come from becomes a question about model calibration. Why does this route keep failing at the same crossing becomes a question about route optimization parameters. Why is Ray's name attached to this pattern becomes a question about association logic and threshold behavior. The operational truth becomes technical language because technical language is what can cross the boundary. When the answer returns, it can be technically precise and operationally useless because the system may be functioning exactly as designed even though the decision remains wrong.

The model is operating within expected parameters. The result falls inside tolerance, the data feed is complete, and no technical defect has been identified. Those answers may all be accurate while failing to answer the question the floor was actually asking. The question was never whether the software was broken. The question was whether the decision was right, and organizations often have stronger mechanisms for investigating the first question than the second.

A failed integration produces evidence. A system outage creates a ticket, an owner, an escalation path, and a resolution process that everyone recognizes as necessary. A broken data feed can be identified and repaired because the failure exists in a category the organization already knows how to manage. A model that works exactly as designed while producing a recommendation that no longer fits the operation is harder to classify. Nothing has failed in the technical sense, which means the strongest mechanisms for forcing attention may never activate. The black box can remain closed because the system has not failed according to the system. The people outside it are left arguing about whether the output should still be trusted, and the burden of that argument tends to fall on the people whose evidence is least formal.

This is where procurement and governance become operational questions even though most people doing the work will never participate in either process. The decision to acquire a system determines more than cost and capability. It can also determine what the organization will later be able to know about the decisions the system influences. A company that intends to place a proprietary system inside a consequential decision process has to ask different questions from a company buying ordinary productivity software. It has to know what happens when the output conflicts with the knowledge of experienced people, what information can be examined, who can examine it, and how quickly that examination can happen.

The organization also has to know whether it retains enough practical authority to stop the output from becoming action while the disagreement remains unresolved. Those are not minor technical provisions because they determine who is actually in control when the system and the operation disagree. The issue becomes more important as AI moves from recommendation into action. A proprietary model that produces a forecast creates one kind of dependency, while a proprietary system capable of using that forecast to initiate another process creates a more consequential one. The more the system can do before a person intervenes, the more costly opacity becomes. The available time for understanding what happened shrinks while the consequences of not understanding it expand.

An organization can tolerate some uncertainty about how a tool generates a suggestion if a knowledgeable person reviews the suggestion before anything happens. The same uncertainty becomes harder to tolerate when a system can change staffing, sequence work, prioritize cases, create flags, or move information into another system before anyone understands why the first decision occurred. At that point, the black box is no longer sitting beside the operation as an analytical tool. It is operating inside the workflow, influencing conditions that other systems and people will later treat as facts. The contractual boundary does not disappear simply because the technology has become central to the operation. In some ways it matters more because the organization may grow increasingly dependent on a capability whose internal reasoning remains partially outside the reach of the people responsible for its consequences.

Dependence changes the conversation in ways that are easy to miss during implementation. Early in a deployment, challenging the system can still be treated as normal adjustment work because people expect problems, configuration changes, and disagreements between the new tool and existing practice. Later, once workflows, dashboards, reporting structures, staffing practices, and performance expectations have been built around the system, questioning it becomes more expensive. The organization is no longer asking whether the tool works. It is asking what would happen to everything built around the tool if it did not. That difference protects the system more effectively than opacity alone.

A deeply embedded proprietary system may become difficult to challenge not because no alternative exists in theory, but because too much of the surrounding organization now assumes the system's continued presence. Processes depend on it, reports depend on it, people have been trained around it, and other systems may consume its outputs. The organization may no longer simply be using the black box. It may have built itself around the assumption that the black box will remain. That dependence can shift power in ways the people on the floor never see. The manager experiences a recommendation as something that is simply there, but behind the recommendation may be years of integration work, investment, process design, and organizational commitment.

The larger that commitment becomes, the harder it is for one contradictory operational signal to compete with it. What looks like a manager challenging one number can actually be a manager challenging the output of infrastructure the organization has already invested heavily in treating as trustworthy. This is another reason local knowledge becomes expensive to raise. The person closest to the work may believe they are simply saying that the route is wrong, the headcount is insufficient, or the performance flag does not describe what happened. From the perspective of the larger system, however, that observation may conflict with decisions made far above the person's role. The worker is not trying to challenge the technology strategy, and the manager is not trying to reopen the procurement decision. They are trying to solve the problem in front of them, but the structure surrounding the technology makes the problem larger than the person raising it intended.

That is why leaders have to remain accountable even when the technology is not fully theirs. Outsourcing a capability does not outsource the obligation to understand what that capability is doing to people, and licensing a model does not transfer responsibility for the consequences of using its output. The organization chose where to place the system. It chose how much authority the outputs would carry, which decisions would rely on them, how quickly the outputs could become action, and how difficult disagreement would be once the system entered the workflow. Those choices remain inside the organization even when part of the technology does not. The contractual relationship may distribute technical responsibilities, but it cannot remove leadership's responsibility for deciding how the system will be allowed to act on people.

This is where I return to the meeting and the question I did not ask. I cannot claim that asking it would have changed the recommendation because I do not know that it would have. The model may have accounted for the configuration change, the historical comparables may have been appropriate, and the recommendation may have been sound for reasons the summary documentation did not make visible to me. That uncertainty is the point rather than a weakness in the example. I was being asked to participate in a decision without having enough access to resolve a question I believed might matter, and the structure of the meeting made moving forward easier than finding the answer. The system had already earned the room's confidence. My uncertainty had not earned the room's time, and the acceptable outcome that followed ensured the difference would never be examined.

Most black boxes survive through ordinary sequences like that. They do not survive because someone stands over them refusing to answer questions. They survive because nothing goes wrong loudly enough to justify the time, cost, and organizational friction required to look more deeply. They become familiar, embedded, reliable enough, and useful enough that unanswered questions begin to look theoretical. Once that happens, the people using the system stop thinking of limited visibility as a current operational risk and begin treating it as an ordinary feature of the environment.

Eventually, one of those questions stops being theoretical. A person is affected, a manager wants to know why, or a leader asks whether a recommendation can be challenged before the consequence becomes permanent. That is when the organization discovers what it can actually see, what it can actually change, and how much of the explanation sits somewhere beyond the people expected to take responsibility for the result. The organization did not only purchase a capability. It accepted a dependency on a system for interpreting part of the work, and it placed that interpretation inside an accountability structure that still lands on its own people.

The contract can define the relationship between companies, but it cannot make the resulting decisions accountable by itself. Accountability still depends on whether the organization retains enough knowledge, authority, and willingness to challenge the system when what the system says and what the operation knows are no longer the same. The black box may sit behind a contractual boundary, but its consequences do not stay there. They enter the building as staffing numbers, routes, classifications, recommendations, flags, and findings that eventually reach people who are expected to execute them and explain what happens next. That is why the responsibility cannot end with the agreement that acquired the technology. It remains with the organization that placed the system inside the decision chain and with the leaders who decide whether a proprietary answer will be treated as information to examine or as authority that no one close enough to the work can meaningfully challenge.

The Comfort of Objectivity

Numbers appear neutral. They present outcomes without ego, without the variability that comes from human interpretation, without the emotional charge that makes certain conversations harder than they need to be. In environments where decisions carry significant consequences and coordination requires a shared language, that apparent neutrality feels not only useful but essential. It is the closest thing modern organizational life has to a ground truth: a reference point that everyone can stand on without first negotiating whose perception to trust.

The appeal is real and not irrational. I have been in rooms where the presence of data stopped an argument that would have been expensive and inconclusive, where a clear number cut through competing narratives and allowed people to move forward. Data protects against certain kinds of leadership failure. It is harder to sustain a fiction in front of clean evidence. It is harder to favor people based on relationship when the numbers show something different. These are genuine contributions, and the organizations that abandoned rigorous measurement in favor of pure judgment did not produce more equitable or more effective environments. They produced different problems.

Objectivity is not the problem. The problem is the comfort of objectivity: the way it reduces the incentive to ask whether the measurement is capturing what matters. Once a metric is established, it carries authority even when it represents only a partial picture of the reality it was designed to describe. The choices that created it, what to measure, how to weight it, what to leave out, become invisible over time. The metric begins to look like a fact of nature rather than a design decision, and at that point, questioning it feels like questioning reality itself.

I have seen teams make decisions from metrics that were accurate in one dimension and blind to trade-offs in another.

The numbers were not wrong. They were incomplete in a specific way that happened to align with what the organization was already inclined to do, which meant the metric confirmed a direction rather than informed it, and the comfort of having data support the decision prevented the harder examination that might have changed it.

There is a version of this that is more uncomfortable still: sometimes the metric is not incomplete by accident. Sometimes what gets measured reflects what someone in a position of authority wanted to see measured, and what gets left out reflects what that same person preferred not to examine. Data launders preference. It converts the priorities of the people who designed the measurement into outputs that appear to have no authors. That conversion is powerful, because it places those priorities beyond the range of ordinary challenge. To question the metric is to question the evidence. To question the evidence is to appear unreasonable.

Objectivity does not remove subjectivity from the system. It conceals it. The choices embedded in what gets measured and how it gets weighted continue to shape every outcome. They just do so from inside the apparent neutrality of a number, which makes them significantly harder to see and significantly harder to change.

The System Is Not Neutral

Consistent is not the same as neutral. A system that applies rules uniformly is consistent by definition. Whether those rules are neutral depends on what they were designed to produce, and what they were designed to produce reflects the priorities of the people who built them. That distinction matters enormously and is almost never the subject of direct conversation inside the organizations the system serves.

I have worked within systems that produced reliable outcomes across all the dimensions they were designed to measure, and that simultaneously created patterns in what they were not measuring that would have been unacceptable if they had been visible. The system was functioning exactly as designed. The issue was not in the execution. It was in the design: in the specific choices that had been made, years earlier, by people who were no longer present, about what the system was supposed to protect and what it was supposed to ignore. Those choices had been validated by consistent outputs ever since, which made them progressively less visible as choices and progressively more visible as the way things are. Every system that calls itself neutral is making an argument. The argument is: trust the outputs without examining the inputs. That argument should be declined, every time, by everyone who has the standing to decline it.

There was a performance ranking system I used for years that sorted workers into tiers based on productivity data. The outputs were clean, updated automatically, and arrived in my review materials already formatted for decision-making. I used them to inform staffing conversations, coaching priorities, and in some cases the determination of whose hours would be reduced when volume contracted. The system had been built before I arrived and would continue running after I left, and nothing about its outputs suggested I should look further than what it showed me. I did not look further. What I understood much later, after I had left that operation, was that the productivity benchmarks the system used as its baseline had been set during a period when the workforce was substantially more experienced than the one I was managing. The model had not been recalibrated. It was measuring my team against a standard that did not describe them, and what I came to see was that the workers still developing the speed assumed by the benchmark were often among those most vulnerable when hours were reduced. The system had not been designed to produce that outcome. It had been designed to optimize productivity. What it had not been designed to ask was who paid the cost of that optimization, or whether the baseline it was optimizing against was still an accurate description of the floor. Those were questions the system did not raise. I did not raise them either.

What the system protected, I came to understand, was not performance. It was the assumption that performance had already been correctly defined. The benchmark was not examined because the benchmark had been producing consistent outputs, and consistent outputs read as validation. For the workers whose hours were reduced, the consequence was not theoretical. Fewer scheduled hours meant less pay. The numbers were accurate. What produced them was not examined. I had interpreted the outputs and acted on the results and told myself that was the job. It was the job. It was also the mechanism by which a design choice made before my arrival, by people with their own priorities and their own incomplete picture of the operation, continued producing consequences I was responsible for and not responsible for in equal measure. That is what embedded priorities look like from inside the system. They do not announce themselves. They arrive as findings.

I accepted it. Not once, in one clearly compromised moment I could point to and examine, but incrementally, across meetings where outputs were presented as findings rather than as the products of choices, where my job was to interpret results rather than interrogate the methodology that had produced them, and where the person asking me to do so was someone whose judgment I had professional reasons to align with. I told myself that examining the inputs was someone else's function, that my role was downstream of the design. That was accurate in a narrow organizational sense and untrue in every sense that mattered. The choices embedded in what the system was measuring had been made by people with the same partial information I had. My not examining them did not make them more neutral. It made them harder to change.

Who Audits the System That Audits You?

Understanding that the productivity benchmark had become disconnected from the workforce raised a question I had not asked when I was using it. We were evaluating workers against that system every week, reviewing their movement, questioning their performance, and acting when their numbers moved outside the expected range. The people were required to keep proving that they still met the standard. What I could not identify was the equivalent process requiring the standard to keep proving that it still described the work.

That imbalance had been hiding in plain sight. Every reporting cycle tested the workers again because their numbers changed, their performance changed, and the conditions around them changed. The benchmark itself entered each new cycle carrying the authority it had accumulated from all the cycles before it. Nobody needed to explicitly declare that the standard was still valid because its continued use had become the evidence of its validity. The system had survived long enough that longevity began to look like verification.

Nothing about the reports warned us that this assumption might be wrong. The calculations still ran, the rankings still populated, the performance bands still looked familiar, and the dashboard continued producing orderly outputs on schedule. There was no technical failure announcing that the conditions underneath the benchmark had changed. The system could remain perfectly functional while becoming progressively less accurate as a description of the operation it was judging.

That distinction matters because organizations are very good at recognizing when people move away from the standard. A worker who falls outside an expected performance band can be identified quickly, a manager whose operation misses a target can be questioned, and a building that begins drifting against its peers can trigger additional attention. Entire systems exist to locate those forms of variance and assign someone responsibility for correcting them. What organizations are often less prepared to detect is the opposite condition: the moment when the standard begins moving away from the people and conditions it was created to measure.

The benchmark does not usually create an alert saying it has become stale. A forecasting model does not announce that the operation it learned from no longer resembles the operation it is predicting. A routing model does not automatically know that the street pattern an experienced driver encounters every afternoon has changed in a way its inputs do not capture. The system continues doing what it was built to do, and the cleaner that process looks, the easier it becomes to treat continued operation as evidence that continued operation is justified.

This is where consistency becomes dangerous when it is confused with validation. A system can apply the same methodology correctly every day and still become wrong about the environment to which that methodology is being applied. Repeating an old assumption with greater precision does not make the assumption current, and producing stable outputs does not prove that the relationship between the model and the work remains stable. It proves only that the mechanism continues to produce what it was designed to produce.

Large organizations need stability in their measurement systems, so none of this means every benchmark should be reopened whenever someone dislikes an outcome. Metrics, models, and standards become useful precisely because people can organize around them without renegotiating their meaning every morning. The problem begins when the need for stability quietly eliminates the obligation to revalidate the assumptions underneath it. At that point, the system becomes something everyone else must continue adapting to while nobody is clearly responsible for asking whether the system should be adapting too.

That is the governance question the outdated benchmark exposed for me. Someone had designed it, someone had approved it, and thousands of later decisions could be made from it without anyone being responsible for deciding when its original assumptions had expired. The workers remained subject to continuous evaluation, while the instrument evaluating them could survive on the credibility of decisions made years earlier. Once I saw that imbalance, the question was no longer simply whether the benchmark had been wrong. The more important question was who should have been checking.

The difficulty is that once a system acquires enough institutional authority, questioning it requires more than noticing that something feels wrong. The person raising the question has to challenge not only the output but the accumulated history behind the output. Prior decisions have been made from it, reports have been built around it, leaders have relied on it, targets have been calibrated against it, and performance conversations may have been conducted for years on the assumption that the benchmark was valid.

The longer a system runs, the more organizational structure accumulates around the assumption that it should continue running. What began as a technical choice becomes part of how the organization understands itself. Examining the original choice can begin to feel like reopening every decision that depended on it, which makes investigating one worker considerably easier than investigating the standard used to judge thousands of workers.

That cost discourages examination even when nobody consciously intends to protect the system. It is easier to investigate why one worker fell below the benchmark than to investigate whether the benchmark should still exist in its current form. The first question produces a manageable answer, while the second threatens to widen until it touches decisions made by people who may no longer be in the room.

So the organization investigates the smaller question. I did that repeatedly by looking at the names that moved, the workers below the expected band, the managers whose teams were drifting, and the locations whose numbers needed attention. The report was telling me where to look, and I followed it because following it was part of the job.

What I failed to notice was that the report could only direct my attention outward. It could tell me which workers deviated from the benchmark, but it could not tell me whether the benchmark had deviated from reality. The instrument was designed to locate problems somewhere else, which meant the instrument itself remained outside the field of view.

That distinction becomes more important as the systems become more sophisticated. A traditional metric can usually be examined with relative ease because you can look at the numerator, the denominator, the target, and the conditions under which the target was established. You can argue about whether the standard makes sense and, with enough authority, change it.

A model is harder because more assumptions sit underneath the output. An agent ecosystem makes the problem harder again because the output of one system can become the operating condition another system evaluates. By the time a visible failure appears, several systems may have acted on assumptions that nobody has independently reexamined.

The forecasting agent produces a number, and the labor scheduling agent treats that number as the best available description of what the day requires. The routing agent works inside the staffing conditions that result, and the monitoring agent evaluates deviations occurring inside an environment the other systems helped create. Every system is capable of checking something, but none of them is necessarily checking whether the others are still right.

The monitoring agent can identify that Marcus is behind his expected route position. It can compare completion times, scan events, and route sequence against the standard it has been given, and it can detect deviation with a precision no manager walking the floor could reproduce manually. What it cannot determine from that deviation alone is whether the route expectation was valid in the first place.

It does not know whether the routing agent had the right local information, whether staffing conditions affected dispatch timing, or whether the definition of normal still matched the conditions under which Marcus was operating. It audits Marcus against the system. It does not audit the system against Marcus's reality.

That direction of scrutiny matters because the system has access to the worker's behavior in extraordinary detail. It can see timestamps, scan records, completion rates, exceptions, sequence changes, response times, and whatever other signals the organization has decided are relevant. Marcus does not have equivalent visibility into the assumptions being used to judge him.

He sees the route, the expectation, and eventually the flag created when the expectation has not been met. The system sees his deviation immediately, while his knowledge that the route itself may be wrong has to travel through a notes field, a manager, an escalation process, or a review cycle before it can reach anyone capable of examining the system that produced it.

The asymmetry is therefore not only informational. It is procedural because one kind of deviation has an automated pathway while the other has a human pathway. Those pathways do not move at the same speed, and they do not carry the same authority.

The system's finding appears instantly as data. Marcus's finding appears later as explanation, which means the system's interpretation has already begun shaping the record before his interpretation has reached anyone with authority to change it. The same organization that can detect a driver's variance in minutes may require weeks to decide whether the route standard generating the variance needs to be reconsidered.

Denise lived inside another version of the same imbalance. Her dashboards could show exceptions faster than her floor walk ever could, and the system could surface workers, stations, and categories that had crossed defined thresholds continuously while she managed everything else. That capability looked like improved oversight, and in a narrow sense it was.

She could see more measurable activity than she had ever been able to see before. The problem was that increased visibility created the impression that the system itself was also becoming more accurate, even though visibility and accuracy are not the same thing. A dashboard can see more of what it has been designed to see without becoming any better at noticing what its design leaves out.

Denise's old floor walk had performed a different function. She was not only looking for known deviations. She was looking for conditions that did not yet have categories, problems that had started before the formal measures changed with them, and situations where the building had begun separating from the official picture.

When the walk disappeared, one of the organization's informal ways of auditing the measurement system disappeared with it. Denise could notice when a station behaved differently even though its metrics remained inside tolerance, hear when conversations had changed before engagement appeared in any report, or recognize that a process was meeting its target only because experienced people were compensating for something the formal system had not identified.

The floor was not just something she managed. It was one of the places from which the system itself could be checked. That made her physical presence valuable in a way the dashboard could never record because the dashboard could only evaluate the operation through the categories it already possessed.

When organizations move more observation into the system itself, they can accidentally remove the independent reference point that would tell them when the system is becoming wrong. The dashboard becomes both the instrument and the evidence that the instrument is working. The cleaner the report looks, the easier it becomes to assume that the process producing it must also be clean.

A scale illustrates the problem more clearly. If a scale is wrong, weighing the same object repeatedly does not make the scale more accurate. It only produces a consistent error, even though every measurement may agree with the measurement before it.

The missing step is comparison against something outside the instrument. Organizations need that comparison too because they need people and processes capable of asking whether the system's description of reality still matches reality before the mismatch becomes large enough to generate a visible failure.

Without that check, repeated output can become circular proof. The model is trusted because its results are consistent, and the results are interpreted through the model because the model is trusted. Each new cycle then becomes another reason not to question the assumptions that produced the last one.

The important question is who is supposed to interrupt that cycle. The answer is often surprisingly unclear because the team that built the system may have moved on to the next deployment, while the analysts maintaining the dashboard may be responsible for data quality but not for whether the underlying performance philosophy is still correct.

Operations leaders may own the outcome without owning the model. Managers may be expected to act on the output while having no authority to change the framework producing it. Everyone owns a piece, but nobody necessarily owns the question of whether the whole thing still describes the work accurately.

That ownership gap becomes more consequential as systems act more independently. A report that is wrong can mislead a meeting, but a system that acts on a wrong assumption can change the operation before anyone recognizes that the assumption has become stale. The difference is not merely accuracy. The difference is timing and consequence.

If the forecasting model begins drifting away from the conditions on the ground, the mistake does not remain inside a forecast presentation waiting for a person to notice it. The number can move into labor scheduling, affect who comes to work, alter floor conditions, shape route execution, and eventually generate monitoring flags tied to individual names. By the time the organization sees enough evidence to question the original assumption, the system has already created evidence that appears to confirm itself.

The forecast reduced staffing, and the reduced staffing created pressure. The pressure produced deviations, the monitoring system detected those deviations, and the deviations were treated as evidence that local performance required correction. From inside the individual systems, nothing necessarily looks wrong because each system can point to the output it received and the rule it correctly applied.

That is precisely the problem. Each system can perform correctly within its own definition of correctness while the combined result moves farther away from what an experienced person looking at the whole operation would call reasonable. The forecasting agent can minimize its error against the historical benchmark, the labor system can optimize against the forecast, the routing system can optimize within the conditions it receives, and the monitoring system can accurately detect deviations from the route.

Every component can pass its own test while the people living inside the combined result know the day does not make sense. A system like that cannot be audited only by asking whether each part completed its assigned function. The audit has to reach the relationship between the function and the reality the function was supposed to serve.

That means asking uncomfortable operational questions before a failure forces them. When was the benchmark established, what has changed since then, and does the workforce the model assumes still resemble the workforce doing the work? Are the conditions that produced the historical pattern still present, and when experienced people repeatedly override or work around a recommendation, is the organization recording those actions as human variance or examining them as possible evidence that the model is wrong?

Those are not abstract questions about whether artificial intelligence can be trusted. Trust is too broad and too easy to argue about. The useful question is whether this specific system, operating on this specific process, with these specific assumptions, still deserves the authority it has today.

The answer can change without the original system having been a mistake. A system can have been right when it was built and wrong now, just as a benchmark can have improved performance for years and become harmful after the operation changes. A routing model can outperform manual planning overall and still develop blind spots that experienced drivers recognize before the model does.

None of those possibilities invalidates the original decision to build or deploy the system. The mistake is treating original success as permanent authorization. Once a system has proven useful, organizations have a tendency to let that proof travel farther into the future than the evidence actually supports.

We do not manage people that way. A manager who performed well five years ago is not exempt from current evaluation because past performance once established competence, and a worker who met the standard last quarter is still measured this quarter. Organizations understand intuitively that human performance has to be reassessed because people, roles, expectations, and conditions change.

Systems should not receive a more durable presumption of competence than the people they are measuring. If anything, they require a different kind of scrutiny because their errors can scale farther before becoming visible. One manager making a bad judgment can affect a team, while a benchmark embedded across hundreds of teams can reproduce the same outdated assumption everywhere while giving each local manager the impression that the result came from an objective standard.

The scale makes the error look less personal, but it also makes the error larger. The organization that audits its people every day while auditing the assumptions judging those people only when the system is first deployed has built accountability in one direction. It knows how to ask whether the worker still meets the standard but has not built the equivalent habit of asking whether the standard still meets the work.

That is the governance gap hiding inside the dashboard. I was part of it because I trusted reports that had been part of the operation longer than I had, that people above me trusted, and that produced outputs consistent enough to feel established. Examining the methodology felt like someone else's function, and inside the structure of the job that explanation made sense.

It was also how the system avoided examination. The people who designed the benchmark were not standing beside me demanding that I preserve it, and the system itself did not need to defend its assumptions. The organization around it had absorbed the idea that methodology belonged upstream while the outputs belonged to those of us expected to act on them.

My job was to interpret the finding and respond to what it showed. Somewhere else, someone else was presumably responsible for asking whether the finding still deserved to exist in that form. The danger is that the person somewhere else may have believed the same thing about someone farther upstream.

This is how systems become durable without anyone consciously choosing to make them permanent. Responsibility for questioning them moves outward until no one standing near the consequence believes the question belongs to their role. The system keeps running, the worker keeps being measured, the manager keeps responding to the findings, and the benchmark survives another cycle because the cycle begins by assuming the benchmark belongs there.

Breaking that pattern does not require rejecting automation, metrics, or algorithmic management. It requires treating the system as something that remains accountable after deployment rather than something whose accountability ended when implementation was approved. A system that evaluates people should generate evidence about itself as seriously as it generates evidence about them.

Overrides should matter because repeated overrides can indicate more than resistance to a tool. Local workarounds should matter because a workaround repeated across experienced people may be evidence that the formal process is missing something. Persistent disagreement between experienced people and model outputs should matter, not because the people are automatically right, but because repeated divergence is information about the relationship between the model and the operation.

The same discipline applied downward has to be applied upward. If a worker repeatedly falls outside the expected band, the organization investigates, and if a route repeatedly requires experienced drivers to correct it, the routing logic deserves investigation too. If managers repeatedly compensate for the same staffing assumption, the assumption should not remain protected simply because their compensation kept the final metric green.

Green can hide a system that is wrong and people who are working hard enough to make it look right. That may be the most important reason to audit the system because the people closest to the work often prevent flawed systems from producing visible failures. They adjust, compensate, double-check, reroute, absorb, and improvise until the operation remains inside tolerance.

Their competence can then become evidence that the system is functioning. The better they are at compensating, the longer the system can remain wrong without being challenged. What appears from above as proof that the standard works may actually be proof that experienced people are spending their judgment and attention covering the distance between the standard and reality.

Eventually that compensation weakens. An experienced worker leaves, a manager stops walking the floor, volume increases, or the accumulated attention cost removes the extra capacity people had been using to close the gap. The failure finally appears, and the organization begins looking for the person or process closest to where the metric turned red.

By then the system that helped create the conditions may already have spent months or years producing green reports. That is why the audit cannot begin with the failure. It has to begin while things still look like they are working, when questioning the system feels least urgent and therefore matters most.

The system is watching the worker, the dashboard is watching the manager, and the monitoring agent is watching the deviation. The organization has built extraordinary capability for seeing when people move outside the boundaries it established. What it has not always built is an equally strong capability for seeing when the boundaries themselves have moved away from the work.

A system should not earn permanent authority merely because it continues producing answers. It should have to remain connected to the conditions those answers claim to describe, and someone with enough standing to matter has to be responsible for testing that connection before the gap becomes somebody else's performance problem.

The people will continue to be audited, the operation will continue to be measured, and the deviations will continue to be examined. The same discipline has to reach the system itself, because yesterday's definition of correct does not become permanently correct simply because the dashboard continues to display it.

The System That Doesn't Wait

The change I want to describe arrived without announcement. There was no meeting where someone explained that the role had shifted, no memo outlining what the organization was now asking people to do differently. What arrived instead was a slow, structural inversion in the sequence of events. Work that used to begin when someone decided to begin it started beginning before anyone had formally chosen its direction.

The morning I noticed it most clearly, I walked into a building where the shift had been running for twenty minutes before the first manager arrived. Not improperly. By design. Routes had populated. Assignments had distributed. Exceptions had been queued for review. The day had already taken shape before a person with authority had logged in, walked the floor, or asked the first question about what the operation actually needed.

That was the part that felt different. The manager who arrived at five-fifteen did not start her day by deciding what to do. She started by inheriting what had already been done. Her first forty minutes were not planning in the old sense of that word. They were recovery: correcting a routing decision that did not match current conditions, reassigning an exception that had been categorized incorrectly, and managing the downstream consequences of a staffing adjustment made before she had seen the floor.

This is the practical meaning of a system that does not wait. It is not simply a system that produces information for people to consider. It is a system that moves the work forward, builds the first version of the day, and then surfaces the pieces humans are expected to correct. That distinction matters because it changes the position of the person inside the process. The traditional role of a manager was to shape what happened before the day hardened around a plan. The newer role is often to monitor, correct, explain, and absorb a process that already has momentum.

You are not steering in the same way. You are monitoring, correcting, explaining, and absorbing. Monitoring is a different kind of work from leading because it begins after something else has already acted. It requires attention, discipline, and judgment, but it does not always carry the same authority as deciding what should happen before the system has made its first move.

The people inside these systems did not volunteer for this repositioning. Most of them could not have named when it happened. It arrived incrementally, through tool updates, process improvements, automated dispatch functions, faster dashboards, and efficiency gains that each looked reasonable in isolation. A dispatch function that used to require three people and a morning planning conversation was automated, first partially, then substantially. The planning meeting became a review meeting. The review meeting became shorter because the decisions had mostly been made.

At no point did anyone stand up and say the job had changed. The title stayed the same. The responsibility stayed the same. The language of leadership stayed intact. But the sequence underneath the role had shifted. The manager no longer entered a day waiting to be shaped. She entered a day already configured by systems that had acted while she was still driving to the building.

Marcus is the right example here, not because his situation was extreme, but because it was ordinary. He showed up each morning to a route that had already been built, a sequence of stops assembled from historical patterns, completion data, package volume, access assumptions, and the categories the system used to define difficulty. He did not plan the route. He drove it. When the route made sense, the system disappeared into the background. When it was wrong, the system became impossible to ignore.

The school zone was the kind of detail a model could easily miss and an experienced driver would know immediately. Marcus knew the crossing guard came out at a certain time, that traffic backed up on clear mornings when visibility changed, and that the sequence on his scanner would put him there at exactly the wrong moment. The route looked efficient from a distance. It looked different to someone who had driven those streets for eleven years.

His knowledge had no useful place to go in the moment it mattered. A notes field existed, and Marcus had used it before, but entering a note about a routing problem routed his observation into a process that ran after the route had already been driven. By the time anyone reviewed the note, the day it could have changed was over. The system was not learning from what Marcus knew in time to use it. It was producing a record of what it had decided, and Marcus was executing inside that record.

That is how the loss of authority often feels from the floor. It does not feel like a machine taking over. It feels like arriving slightly late to your own work. The sequence is there. The expectation is there. The timer is running. The exception rules are already defined. What remains for the person is not to decide the day, but to perform inside the day the system has already built.

The same thing happens to managers, only at a different level. The staffing expectation is already populated. The route structure is already visible. The exceptions are already waiting. The dashboard has already decided what deserves attention before the manager has had the chance to decide what the building actually needs. The first task of the day becomes reconciling the system's version of the operation with the operation itself.

There is still judgment in that work, and that matters. The manager correcting the morning's exceptions is exercising real judgment. Marcus, building a parallel sequence in his head while following the route on his scanner closely enough to stay within the acceptable band, is exercising real judgment. The issue is not that judgment disappears. The issue is that judgment gets repositioned. It becomes corrective rather than originating. It acts at the edges of what the system has already decided instead of shaping the decision before it takes effect.

That repositioning changes people. It teaches them to wait for what the system has produced before they decide what they are allowed to think. It teaches them to treat their own knowledge as supplemental, useful only after the official version has appeared. It teaches managers to ask what has been flagged before asking what is actually happening. It teaches workers to adapt to the route, the schedule, the rate, and the exception category before asking whether any of those things describe the reality in front of them.

The organizations that have made this transition are not, in most cases, organizations that thought carefully about what they were doing to the people inside them. They were solving specific operational problems: reducing manual steps, improving consistency, addressing variation, and moving decisions faster than individual managers could move them. The systems they built achieved many of those things, and that is part of why the transition was so easy to accept.

What those systems also did, as a consequence that was never fully calculated, was absorb the initiating intelligence of the people who had been doing that work. The reading of conditions, the allocation of resources, the sequencing of priorities, and the judgment about what needed attention first began migrating into the system's logic. What remained for the person was the management of exceptions the system had already defined, in categories the system had already established, through processes the system had already configured.

The title on the badge still said manager. The actual function was becoming something closer to steward of a process that belonged to the system. That stewardship required competence, discipline, and judgment, but it did not carry the same authority as deciding what the process should be before it acted. The person remained responsible for the day, but the first version of the day increasingly belonged to something else.

This is not yet the full agent ecosystem the later chapters will examine. It is the earlier condition that made that ecosystem possible. Before routing agents, forecasting agents, labor scheduling agents, and monitoring agents began shaping the same operation through their combined consequences, organizations had already accepted the premise that the system could act first and people could correct later. Once that premise became normal, the rest followed more easily than anyone wanted to admit.

A system that operates before anyone is watching is also a system whose early decisions are not being made by the people who will later be held responsible for them. The exceptions it queues, the corrections it misses, and the situations it processes through a category that does not quite fit are the places where human judgment used to enter earlier. Those gaps have not disappeared. They have become harder to see because the system fills them in before anyone thinks to ask whether the answer is right.

The responsibility for those consequences does not resolve cleanly because there is no single point where the failure can be isolated and assigned. Everyone in the chain made a reasonable decision given the information and authority available to them. The system did what it was configured to do, the manager corrected what she could, and Marcus drove the route he had been given while adjusting where the conditions required it. No single moment looked like a failure because each individual step was defensible on its own. The work continued, which is often enough for the organization to conclude that the process worked.

What is harder to see is that the sequence itself had changed. The system was no longer waiting for someone to examine the conditions, decide what mattered, and then set the work in motion. It had already built the route, distributed the assignments, established the expectations, and created the categories through which whatever happened next would be interpreted. By the time the manager arrived, the important questions were no longer being asked in open space. They were being asked inside a structure that had already produced an answer and begun acting on it.

That is why this change is so difficult to name while it is happening. Nothing announces that authority has moved, because the system does not need to remove anyone's title, rewrite anyone's job description, or declare that human judgment matters less than it did before. It simply begins earlier. It starts the work, builds the record, queues the correction, and leaves the person to enter the process after the first decisions have already been made. The manager still appears to be managing and Marcus still appears to be making choices, but both are operating inside conditions that existed before either of them had the opportunity to shape them.

Over time, that timing difference becomes a form of authority. Whoever acts first defines the starting point, and everyone who follows is forced to respond to what is already there. A manager who disagrees with a staffing decision is no longer proposing a plan; she is asking the organization to undo one. Marcus who knows the route is wrong is not designing a better sequence; he is deviating from the sequence the system has already designated as correct. The burden therefore shifts quietly from the system having to justify its first decision to the person having to justify why that decision should be changed.

From above, this can still look like efficiency. Work begins sooner, fewer decisions wait for human attention, and the organization can move at a speed no individual manager could sustain manually. From inside the operation, however, the experience can feel entirely different. It can feel like arriving in the middle of a sentence someone else started writing before you walked into the room, then being held responsible for how the sentence ends. The system does not have to eliminate human judgment to change its role. It only has to act early enough that human judgment becomes the correction rather than the beginning.

The Weight of More

The case for automation, inside every organization that has invested in it seriously, rests on a promise about capacity. Free up the manual process and you free up the people who ran it. Automate the repeatable and you release human judgment for the things that actually require it. The argument is intuitive, and in certain narrow senses true. The processes that were automated did consume time. When the time was recovered, something real was gained.

What the argument does not account for is what the system does with its own output. Because the system does not just do the work. It documents the work, at a level of granularity that was not previously possible, and it generates information about the work that did not previously exist, and it surfaces that information to the people who are now responsible for what the system is doing. The monitoring layer that replaces the manual process is continuous. It does not end when the process ends. It runs alongside everything else, alerting, flagging, reporting, requiring attention in volumes proportional to the system's own capability. The capacity freed by automating the task is, in many cases, immediately consumed by managing the information the automation generates.

I watched this happen in an operation that had just deployed a new workflow management tool designed to reduce the manual coordination that moved work through a processing center. The tool worked. Coordination time dropped. Errors in handoffs decreased. The people whose job had been to manage those handoffs now had recovered time in their day. What filled that time, without anyone explicitly deciding it would, was the information the tool produced. The tool generated a dashboard. The dashboard surfaced metrics that had not previously been tracked. The metrics created accountability for conditions that had previously been managed informally. The accountability created reporting requirements. The reporting requirements created meetings to review the reports. Within a quarter, the time recovered from the coordination process had been substantially reinvested in managing the visibility that replacing the coordination process had created.

This is the overwhelm that automation produces when the full cycle is examined. It is not a failure of implementation. It is a feature of how these systems are designed, though not one that appears in any of the design documents. The system's value is measured in efficiency metrics, and those metrics do not capture what gets added to the roles of the people who oversee it. The dashboard that replaced the morning count now runs all day, surfacing conditions the old process would have identified once per shift. The alert system that replaced the manager's floor walk now generates responses at the same rate the manager's judgment once absorbed through direct observation. More is known. More is responded to. The response is constant. The person doing it does not have more hours in the day.

Denise, whose forty-minute floor walk had been absorbed into the expanding administrative compliance workload, did not stop walking because she had been replaced. She stopped walking because her attention had been redistributed. The walk had taken forty minutes. The laptop required forty minutes. When the reporting compliance scores began appearing in her monthly review, the allocation became clear: the laptop was the scoreable activity. The walk was not. She made the rational choice. The operation lost the kind of knowledge that only a floor walk produces. The system gained the compliance score it was configured to reward. Nobody recorded that trade in any ledger. It did not appear as a cost anywhere. It appeared, months later, in a customer complaint that was attributed to human error at the sort stage, and assigned a corrective action that was a retraining reminder, and closed.

The pressure to optimize does not settle once the tools are deployed. The system identifies the next gap, because identifying gaps is what it was built to do, and then it surfaces the gap to the person whose job includes closing it. That surfacing is continuous. The next gap appears as soon as the last one is addressed, sometimes before. The person responsible for the gaps is being presented with information about all of them at a rate that bears no relationship to their actual capacity to address them. The result is not a workforce that is more productive because the manual work has been automated. The result is a workforce that is differently exhausted: not from the physical weight of tasks that have been removed, but from the sustained, ambient pressure of monitoring a process that is always running and always surfacing something that needs a response.

What this produces over time is the inability to disengage. Not as an individual failure of discipline, but as a structural condition of the role. The system does not have operating hours in the traditional sense. It runs through the evening and over the weekend and during vacation, generating alerts, queuing exceptions, updating forecasts, and refreshing the metrics that define how the operation will be judged. Managers in the operations I knew were required to have the company's operational app on their phones. The app provided access to live dashboards, exception alerts, forecast changes, staffing signals, aging work, missed targets, and performance categories from anywhere. That access made round-the-clock awareness possible, even when no policy explicitly required an immediate response to every alert. It allowed a manager to see the operation without being physically inside the operation, which was useful. It also meant the operation could follow the manager into places where work used to stop.

The phone in the manager's pocket was not a convenience. It was an extension of the monitoring obligation into the hours when the organization used to be absent. Nobody had to explicitly say that every alert required an immediate response after hours. Nobody had to write a policy explaining that a manager should glance at the app before dinner, during a weekend, or while supposedly taking time off. The system simply made constant awareness possible, and in these environments, possibility has a way of becoming expectation without the expectation ever being stated. There is an undocumented toll in that kind of awareness. It does not appear in labor planning. It does not show up in productivity reports. But it accumulates in the manager's nervous system, in the quiet inability to stop checking, stop anticipating, and stop wondering what changed while they were trying to recover.

One manager who had access to the new consolidated dashboard, across eight performance categories updated in near real time, told me she did not know when she was supposed to look at it. She had a morning standup, a midday check, an end-of-shift review, and the expectation, never stated but thoroughly understood, that she would respond to alerts as they came through. The dashboard had not replaced her previous workload. It had added a new layer to it, a layer without defined boundaries between on-time and off-time, because the system did not distinguish between the hours when she was supposed to be paying attention and the hours when she was supposed to be recovering.

When I asked her whether she felt supported in managing that boundary, she looked at me the way experienced people look at a question that does not quite map onto the world they are living in. She said she managed. I believed her. I also knew what managing meant in that context, and what it was costing her to do it.

The efficiency was real, and the exhaustion was also real. Both were built into the same design. The organizations that count only one of them are not accounting for the actual cost of what they built. That accounting will eventually arrive, not through a formal review, but through the accumulated evidence of what happens to organizations that have been quietly withdrawing from a resource they did not think to measure.

The Only Thing Left to Give

There is a precise way to describe what the system actually consumes. Time is part of it, but time is the surface. Effort is closer, but effort still implies something voluntary, something that can be withheld and restored. What the system consumes, at the level that matters most, is attention. The capacity to be present to a situation: to notice what is happening, to hold multiple conditions in view at once, to respond to what is most important among competing demands. Every dashboard, every alert, every monitoring layer, every continuous process is drawing from this resource. And unlike the output of any automated system, attention does not scale.

This is the constraint that the organizations building these tools have not yet learned to describe. They have solved the labor problem, or at least transformed it: they built systems to extend what labor could accomplish, to automate the repeatable, to coordinate at scale. The systems succeeded. Labor, as a bottleneck, looks different now than it did twenty years ago. The bottleneck has migrated. It has moved from what people can do to what people can hold in their minds at one time, and it continues to move even as the systems that created the shift keep adding capability.

Every chapter of this book has been describing this constraint without naming it directly. Denise stopped walking the floor not because she ran out of capacity to walk it. She stopped because her attention had been redistributed, away from the floor she used to read by feel and toward her laptop that required compliance. The regional leader who spent two weeks accumulating informal intelligence before the planning session had the attention necessary to see what the model was missing. The organization had built a process that ensured her attention could not reach the decision at the moment the decision was being made. Marcus stopped doing his second check not because the check was difficult, but because the throughput pressure had created a competing claim on the same forty-five seconds, and the competing claim came with a metric attached and the check did not. These are not separate stories about separate people. They are the same story, repeated across different roles and different buildings, about what happens when the demands placed on a finite resource keep growing and nobody is responsible for tracking the deficit.

The leaders who fail to push back against the model, which this book has been examining across several chapters, are not failing because they lack courage in the abstract. They are failing in moments that arrive at the end of sequences that have been drawing on the same resource continuously. The meeting where the model's recommendation sits at the center of the table is not the first demand that day on the attention of the people who might challenge it. It is the latest one. The willingness to spend the cognitive and relational cost of dissent, to say out loud that the model has missed something important and to stand behind that claim through the resistance that follows, is itself an expenditure of attention. Organizations that exhaust the people who might make that expenditure should not be surprised when it is unavailable in the moment it would matter.

For most of organizational history, the scarcest resource was labor. Not enough people to do the work. Systems were built to extend what labor could accomplish, to automate the repeatable, to coordinate at a scale that exceeded what individual attention could reach. Those systems succeeded, and the success changed the constraint. Capability is no longer the limiting factor in most of these environments. The people inside them are capable. They have been selected for it, trained in it, evaluated against it for years. What is limited is not what they can do. It is what they can attend to, simultaneously, under sustained load, across the continuous flow of demands the system generates. The system scales. Attention does not. Every time a new capability is added, a new claim is placed on the same finite resource. The ledger does not balance, and the organizations keeping the ledger are not tracking the column where the deficit accumulates.

I was one of the people adding to the ledger. When the consolidated dashboard was proposed, I supported it. The efficiency argument was compelling and the visibility argument was compelling and I believed both of them, and I was also conscious that supporting the expansion was the kind of contribution that registered well in the organizational context I was operating in. When the dashboard went live, I received a briefing on adoption metrics: how many managers were logging in, how frequently, at what hours. Adoption was high. I noted that as a success. I did not think, in that meeting, about what the high adoption rate meant for the people whose jobs had just absorbed a new continuous monitoring obligation, or about what they were no longer attending to in the time that logging in was consuming. I was looking at the metric. The metric said people were using the tool. What the tool was costing them did not have a metric.

What this looks like in practice is a workforce that is technically performing while operating at a fraction of its actual perceptive capacity. The manager is watching six screens instead of walking the floor she once knew. The manager is processing alerts in the gaps between meetings, which are not really gaps because the processing leaves no margin for the kind of thinking that happens only when there is space for it. The executive is reviewing model outputs with the portion of his attention that is not already committed to the reporting requirements that precede this meeting and the communications that will follow it. The decisions being made in these rooms are being made by people whose attention has been distributed across more than any single person can fully hold, and in the gap between what they are carrying and what the decision requires, things are being missed that nobody is positioned to name.

The site manager tracking in her notebook what the system could not see was not doing something heroic. She was doing something ordinary: she was paying attention to what was actually happening in the building she was responsible for. That the organization could not receive that attention, could not route it to a decision, could not act on it before it became history rather than warning, was not a failure of her judgment. It was a failure of organizational design. The building had the information it needed. The people with authority to use that information did not have the attention required to find it inside the formal channels they were monitoring, and the person who held it had learned through experience that the formal channels were the wrong place to put it.

The organizations that will navigate the next phase of this are not the ones that add more tools. Adding more tools adds more claims against the same resource. The ones that will navigate it are the ones that begin to account for attention the way they account for any other scarce input: by understanding where it goes, what it costs to sustain, what the return is on each competing demand, and whether the system being built is consuming the resource faster than the people inside it can replenish it. That accounting does not exist yet in most of these organizations. It does not produce a metric. It does not appear on a dashboard. It accumulates in the quality of decisions made by people who are doing their best inside a system that has no language for what it is asking of them.

What the Room Already Knows

The conversations in the parking lot were always more accurate than the meetings that preceded them. This is not a trivial observation. It is a diagnostic. When the real analysis of a situation happens outside the official container designed to hold it, the organization has not failed to communicate. It has built conditions in which honest communication carries a cost that the formal process does not absorb. The truth is still present. It has simply relocated to the only space where saying it does not require paying for it.

Margaret has been on that floor longer than Terrence has been managing it. She knows things about that operation that do not appear in any report and that nobody with authority over the floor has formally asked her about: which induct stations run tight on Tuesdays when returned undelivered volume spills into the morning sort, which coworkers go quiet in a way that means the next two shifts will need watching before anything shows in the error counts, and where the east loading dock scanner loses the angle on packages above a certain height and requires a manual confirmation the system does not prompt for. This is not incidental knowledge. It is the kind of operational understanding that takes years to build and that disappears, floor by floor, when the people who hold it conclude that the formal structure has no use for it.

She reached that conclusion through accumulation rather than a single moment, though the standup meeting where she asked Terrence where the time was supposed to come from was part of it. That question was an act of surfacing what she knew: the new pick rate expectation did not fit the current headcount when you ran the numbers against the actual conditions on the floor, and she had been running those numbers in her head all morning. The answer she received was the expectation restated in different words. The meeting moved on, the question was not revisited, and what she learned from that exchange was not that the question was wrong. It was that the formal pathway for her knowledge had a particular shape, and the shape did not include a place where what she understood about the floor could change a decision that had already been made at a level where the floor was not visible.

The whiteboard in the break room was the next pathway she tried, and she used it for longer than the standup would suggest. She added to it for months: the recurring problem with damaged freight coming off specific trailers, the east loading dock scanner issue, the Tuesday station backlog, and the weight discrepancies she caught on the high-value sort. Her name appeared on the whiteboard more than anyone else's during that period, not because she was struggling, but because she was paying attention to things the formal system had no category for and was still, at that point, trying to surface them through a mechanism the organization had provided.

What the whiteboard generated, across those months, was silence. Not hostility. Not disagreement. Silence, of the specific kind that accumulates when observations routed to a formal channel produce no visible response. She kept adding to it until she understood that the whiteboard was not a pathway. It was a record of what the floor knew that nobody was reading.

After that, she stopped adding things. She stopped pausing at the east loading dock station for the scanner angle gap. She stopped flagging the weight discrepancies. The knowledge did not go anywhere. It remained where it had always been, in the specificity of her attention moving through that floor every shift. What changed was only what she did with it. The cost of deploying it had come to exceed the cost of keeping it to herself, and she made that calculation the way experienced people make calculations about organizational environments: quietly, without drama, and without the people above her knowing it had happened.

The gap between what organizations officially know and what the people inside them actually understand is not produced by dishonesty. Most people in these environments are not strategically withholding. They are making rational calculations about the cost of surfacing what they know in the contexts where what they know would be actionable. Marcus understood that flagging his route variance would generate a deviation log rather than a conversation about whether the routing was right. Denise understood that the things she caught on the floor walk, the signals that existed before they became logged problems, had no format the formal data layer could receive. Margaret understood that the whiteboard was a record nobody was reading and that the standup was a ceremony for expectations that had already been set. The truth they all held was real. The pathway for it to reach a decision was not.

Truth does not disappear when the formal channels fail to receive it. It moves. It moves into the manager's two weeks of informal intelligence that cannot reach the planning meeting in time to affect it. It moves into the whiteboard filling up with what the floor knows and the structure does not. It moves into the driver who stops entering alternate sequences in the notes field because the system flags him for route variance, not for being right. Each movement tells the same story: the organization still contains the truth, but the truth has learned where it can survive.

Each of these is the same phenomenon: an organization that possesses accurate, actionable information it has designed itself to be unable to use. Margaret is the clearest version of it here. She has been on that floor longer than anyone managing it, and the knowledge she holds is exactly the kind the formal data layer trails by days, weeks, or the full interval between when a problem begins and when it becomes visible in the metrics. The organization has given her every available signal that the cost of surfacing it is hers to absorb alone.

I was in a post-mortem once for a failure that had been visible, from inside the building, for weeks before it materialized. The formal analysis was thorough. It traced the contributing factors through the data, identified the point of origin in a process gap, and produced a corrective action that addressed the documented cause. What it did not do, because it had no mechanism to do it, was ask the people closest to the work what they had known and when they had known it. That question was not part of the process. The process was designed to identify what the data layer had failed to capture. It was not designed to examine what people inside the operation had already understood but had found no usable pathway to surface.

The manager who had seen the problem building, who had a specific and accurate mental model of why the failure was coming and roughly when it would arrive, sat in the root-cause review and said nothing about it. The process was not asking. She had learned, through prior experience in that organization, what happened to observations submitted through informal channels. They routed to a review cycle, and the review cycle ran after decisions had already moved. She had been right about the problem. She was also rational about the cost of having been right too early to have it matter. The corrective action addressed a documented cause. The actual cause, which she understood precisely and could have described in detail, was never discussed.

I was running that review. It was a late-dispatch failure, the kind that surfaced as an outlier in the metrics dashboard and required a formal root-cause review. I had been asked to lead the corrective-action process, and I followed the process, which meant I followed the data. The numbers showed the late departures. The route list showed which waves had missed. The dispatch records showed late staging, grounded vehicles, delivery trucks not ready for departure, overflow packages being sorted on the loading bay floor, and associates pulled off task during the middle of the shift.

The documented cause was real. Nothing in the review was fabricated. But the process pointed me toward what could be measured, named, assigned, and closed. It did not point me toward the people closest to the floor who could have explained what had been building before the miss became visible in the dashboard.

I knew who was in the room. I knew which floor leads had been closest to the failure. I knew which staging supervisors had watched the lanes back up, which associates had seen the staging process collapse, and which drivers had been standing beside their vans waiting for racks that were not ready. I did not ask all of them what they had known. I did not ask because the process did not have a step for that question, and inserting a question the process did not include would have required me to justify the deviation. That deviation would have introduced uncertainty into a timeline the organization needed closed.

So, I produced a clean corrective action. The formal corrective-action record had root causes, owners, dates, and follow-up actions. It looked complete because, by the standards of the system, it was complete. The documented cause was real. The actual cause, the one the people closest to the floor could have described in detail, was only partially addressed. The system got the version of the problem it was built to recognize. The operation kept the part of the problem that did not fit cleanly inside the form.

What that organization contained, distributed across the people doing the work, was something more accurate than anything its formal systems were producing. This is the condition most large organizations are actually in, though they have not built language to describe it. The people closest to execution hold a collective understanding of current conditions that the formal data layer trails by days, sometimes weeks, and sometimes by the full interval between when a problem begins and when it becomes visible in the metrics. The mechanism for moving that understanding from where it lives, in informal conversations, break room whiteboards, and exchanges that happen after the meeting, into a form the organization can act on, is almost never adequately built.

What is missing is not another report. The organization already has enough reports, and in many cases it has too many. What is missing is a way to turn distributed judgment into usable signal before failure hardens into a metric. The tool for that cannot be another top-down forecast that simply processes more historical data and calls the output prediction. It has to work in the opposite direction: from the people closest to the work toward the people with authority to act.

That is what I mean here by a prediction agent. Not a forecasting agent that tells the floor what the model believes will happen, but a structured mechanism for collecting what people close to the work already believe is happening and aggregating those judgments into a signal leadership can see. Its value is not that it replaces judgment with another model. Its value is that it gives judgment a pathway before the system turns the failure into a metric.

The concept that addresses this directly is not new, though it is under deployed in the contexts where it would matter most. Prediction agents used inside organizations are not primarily about forecasting. Their function is like a structured listening system, collecting informed judgments from people close to the work and turning those judgments into patterns leadership can see. They are about surfacing distributed knowledge that the formal pathway cannot reach. When individuals can register their actual assessment of a situation, anonymously and in aggregate, the organization gains access to the truth it already contains but cannot otherwise see.

Margaret, without putting her name on anything and without defending her observations in a standup that is not designed to receive them, could register what she believes about which stations will backlog on Tuesday and which workers are struggling before it shows in the data. The person who thinks the staffing model is undercounting for the current team composition could register that belief without the personal cost of a formal escalation that will not reach the current decision in time. The experienced driver who has been watching the routing sequence send people into the school crossing at the wrong time for three weeks could surface that knowledge before it becomes a logged complaint. The point is not that every observation would be correct. The point is that the organization would finally have a mechanism for seeing what its people believe before the system turns the failure into a record.

That distinction matters. One person's concern may be noise. The same concern appearing across ten people, three shifts, and two locations is not noise. It is a signal, and it may be the earliest signal the organization is capable of receiving if it builds the mechanism to receive it. The purpose of a prediction agent is not to create another dashboard to worship. Used badly, it would become exactly that, another metric stripped of context and used to pressure the people who supplied it. Used properly, it becomes a way to ask the organization what it already knows before the official record catches up.

The reason organizations do not build this more often is not technical. The tools exist. The concept is well understood. The reason is cultural and structural. An organization that builds a systematic mechanism for surfacing what its frontline people actually believe is implicitly acknowledging that what its models report and what its people know are not reliably the same thing. That acknowledgment is uncomfortable for the people whose authority rests on the models. It challenges the premise that the formal data layer is sufficient. It suggests that a planning session built on validated model outputs is operating with less information than someone who has been on the floor for six years and has stopped adding things to the break room whiteboard because nobody was reading it. Organizations in which authority tracks upward with access to data do not readily build mechanisms for demonstrating that the data is incomplete in ways the people below can see and the people above cannot.

There is also something in the existing accountability structure that actively resists it. A prediction agent produces assessments shaped by proximity rather than hierarchy, but those assessments still move inside organizations shaped by incentives. Honest assessments sometimes contradict official projections. Official projections are, in most organizations, the record against which leaders are evaluated. A mechanism that routinely demonstrates that the informal knowledge of the people doing the work diverges from the official projections of the people managing them makes visible a gap that most organizations have a structural interest in keeping invisible. This is not cynicism. It is an accurate description of the incentives in place, and understanding it explains why the mechanism does not get built even when the need for it is obvious to anyone who has spent time in the parking lot conversations.

The organization may say it wants truth, but the kind of truth that arrives before the model confirms it creates exposure for the people who approved the model, defended the projection, and organized the meeting around the official version of events. That is why the silence can become easier to live with than the listening. Listening would require the organization to admit that the people closest to the work are sometimes holding a more accurate picture than the systems designed to describe the work.

The information the organization needs is already present. It was in the whiteboard entries that stopped generating responses. It was in the question Margaret asked in a standup meeting eight months ago that the meeting was not built to answer. It was in the east loading dock scanner gap she stopped pausing for because pausing cost time, and the time had a metric attached to it while the concern did not. The question organizations dealing with this will eventually have to answer is not whether the truth exists inside their walls. It does. It exists in the people who have been on the floor the longest, who know the most, and who have learned through ordinary accumulating experience that the cost of surfacing what they know is theirs to carry alone.

The question is whether those organizations are willing to build the conditions under which that knowledge can reach a decision, or whether they will continue investing in more sophisticated data systems that already have less accurate information than the people working inside them. That choice is not a technology decision. It is a decision about what kind of organization you are willing to be. It is made in ordinary rooms, by people with the authority to choose something different, who have so far found the cost of listening higher than the cost of the silence that replaces it.

What Leadership Actually Means Now

The meeting where the staffing numbers came down was a regional operations review, two weeks before the peak holiday volume window, and Terrence had been watching for it in the way you watch for something you already know is going to be wrong. The recommendation appeared on the screen formatted as a projection, numbers in columns, a staffing plan for the period built from prior-year data and volume benchmarks and the kind of facility-level assumptions that look accurate from the level where accuracy is measured in aggregate. He looked at the numbers. He looked at the room. He looked back at the numbers and did the math he had been doing in his head for six weeks, since before he understood that the number would eventually arrive in a room like this one.

The model was understaffed by three positions for what the period actually required on his floor. Not understaffed by three positions for what the period looked like in the data, which was built on a facility that had since changed its induct configuration and on a team composition that had shifted when two of his experienced leads left due to stress related reasons. Understaffed for the fourteen people he actually had: Margaret, who had been compensating since October for the tunnel scanner calibration gap on the east loading dock that the maintenance ticket had not resolved, and three hires who were eight weeks in and running at roughly sixty percent of experienced rate on the sort lanes that carried the most volume during the 1am to 4am window. The model had his headcount. It did not have any of the rest.

He had been tracking the gap in the way floor managers track things that are not yet in any system: in the timing of breaks adjusted to keep certain stations from running short, in the informal sequencing he and Margaret had worked out to cover the east loading dock without creating a bottleneck downstream, in the number of times per shift he had moved people off their primary assignments to absorb a problem the staffing plan had not accounted for and would not account for until the problem showed up in the metrics. He had six years of that floor. He understood it in a way the projection on the screen did not. The question in the room was whether any of that understanding had a path to the decision being made.

The accountability math is not complicated and most people who have worked in these environments can run it in the time between a slide advancing and the next person in the room starting to speak. Terrence had been in enough rooms to know it without performing the calculation consciously: his development conversation was scheduled for late January, the person presenting the numbers was two levels above him and had been the strongest voice in the prior-year review that had moved him from eleven people to fourteen, the expectation of the room was a confirming discussion rather than a corrective one, and the question he would be asked if he raised his concern, whether he had data to support a different staffing picture, was a question he could not answer in the form the room would require. What he had was six weeks of informal compensations and a maintenance ticket that was still open. What the room had was a validated model output.

What leaders are actually protecting when they defer to a model output instead of acting on what they know is not abstract and is not, in most cases, cowardice. It is the specific and personal architecture of what happens next. Terrence was not protecting himself from being wrong. He had high confidence in what he knew about his floor. He was protecting the relationship with the person who had advocated for his team's expansion, the development conversation in January, the possibility of the next role he had been working toward for two years. A floor manager who challenges a validated staffing model, argues for a different number, and turns out to be right earns a limited form of credibility that the organization moves past quickly. A floor manager who challenges a validated staffing model and turns out to be wrong becomes the person who bet against the process in a regional operations review and created friction that slowed down a decision. That asymmetry is not subtle. It is the first thing experienced people feel when the moment arrives, and it does not take long to calculate, and the answer it produces almost never favors the person who knows something the model does not.

He did not raise it. He asked one clarifying question about the volume weighting for the induct lanes, which was a real question and also the kind of question that shows engagement without introducing uncertainty into a direction that had already been set. The room moved on. The staffing plan was confirmed at the number on the screen. He sat with what he knew about the gap and what that gap was going to cost his team across the next six weeks, the coverage he would not have, the informal compensations that would become less informal as volume climbed, the conversations he would have with Margaret about how to hold the east loading dock without the maintenance issue resolved and without the bodies the model had decided they did not need.

The most honest thing to say about that moment is that nothing in it was irrational. He made the calculation correctly. The organization had built the conditions in which the calculation came out the way it did, in which surfacing what he knew was more expensive than absorbing what he knew, in which the most professionally sound choice was also the choice that kept accurate information from reaching the decision it would have changed. That is not a failure of Terrence's character. It is a function of what the organization has made rational, and the people inside these environments who understand what is happening are not making worse decisions than the people who don't. They are making the same decision with more clarity about what it costs.

What gets protected is not the organization's ability to make a better staffing decision. What gets protected is the individual's ability to move through the next performance cycle without a traceable mistake attached to their name. Terrence's team will absorb the gap. Margaret will compensate for the east loading dock through the ten-to-two window and the system will log her throughput as a deviation and Terrence will absorb the coaching conversation about her numbers without having a way to explain what the numbers are actually measuring. The metric will record the performance of his floor against a staffing model that did not account for what he had been watching for six weeks, and the gap between those two things will settle, as it always does, on the people closest to the work.

The leaders who are doing something different inside these environments share something that does not appear in any review. They have held on to the distinction between what the model knows and what the situation requires, and they have found ways to make that distinction legible inside rooms that are organized to make it inconvenient. They still execute within the structure. They deliver the results the model is tracking, work inside the processes that exist, and understand that every disagreement cannot become a battle over first principles. They are not rebels or chronic dissenters. What separates them is that they have refused to let the confidence of the system replace their own obligation to pay attention.

That sounds simple until you consider what the environment asks of them. The model arrives with data, history, institutional approval, and the appearance of precision. The leader arrives with something less formal: what happened on the floor yesterday, what experienced people are saying, what has changed since the assumptions were established, and what does not feel right even though the dashboard has not yet moved. Holding on to that knowledge requires discipline because the system continuously teaches people which forms of evidence are easiest to defend. The leader has to resist the temptation to treat what is measurable as automatically more real than what is directly observable.

The best leaders I have seen inside these systems are not simply better at challenging them. They are better at translating what the system cannot see into a form the organization is capable of receiving. They know when operational knowledge needs numbers, when a pattern needs examples, when a concern needs to be raised before the meeting rather than during it, and which person has enough standing to keep an inconvenient observation alive long enough for it to matter. They build credibility around that process so that disagreement does not immediately get interpreted as resistance. None of this guarantees the system will change, but it gives reality a path into the decision before the decision hardens.

That pathway is narrow, and it has to be cultivated deliberately. It depends on relationships, timing, trust, and an understanding of how authority actually moves through the organization rather than how the organizational chart says it moves. In healthy environments, leaders create that pathway for the people below them. They make it possible for a manager to say that the staffing model is wrong without first having to prove a failure has already occurred. They create room for someone like Denise to say that the east loading dock has a problem the metrics have not captured yet. They allow someone like Margaret to explain where the time is actually going before the only acceptable evidence is a missed target.

That is why Terrence matters. He did not fail because he lacked judgment, courage, or understanding of the work. He knew the staffing expectation did not fit the conditions on his floor, and he understood why Margaret's question mattered. What he lacked was a pathway through which that knowledge could reach the current decision with enough authority to change it. The room was not organized to receive what he knew, the process was not moving on the same clock as the operation, and the cost of forcing the issue belonged almost entirely to him. Asking him to be more courageous would have missed the point. Leadership should not depend on one manager repeatedly accepting personal exposure just to make operational reality visible.

What leadership actually means now is building structures where people do not have to choose between alignment and accuracy. It means preserving a place for judgment before the system has already converted a condition into a finding, and making sure the people closest to the work can challenge an assumption without being treated as the source of friction. A leader cannot personally inspect every route, every station, every staffing plan, or every model output. But a leader can create an environment where the people who do know those things have a credible way to change the decision before the cost reaches someone farther down the chain.

The organizations that cannot see the difference between leaders who function efficiently inside the system and leaders who actually lead are losing things they do not know how to measure. They lose the six weeks of operational understanding that never reached the staffing decision. They lose the warning embedded in the east loading dock gap before it becomes a root cause entry. They lose sight of the fourteen people on a floor where management is doing almost everything the organization asked correctly, except for the one thing the model has made too expensive to say. Those losses do not appear immediately because the work continues, the numbers still populate, and the system keeps moving. By the time the loss becomes visible, the organization often calls it a performance problem instead of recognizing it as a leadership failure that began much earlier.

Where Responsibility Still Exists

The system narrows the range of available choices. It does not eliminate choice. In the space between what the system defines and what the moment actually requires, decisions are still being made by people, and those decisions still belong to the people who make them, regardless of whether the system has a category for them.

I have made decisions that aligned perfectly with the system and required very little personal exposure. The data supported them, the process endorsed them, and if anything went wrong the accountability had been distributed broadly enough that no single person bore very much of it. I have also made decisions that required accepting that the outcome might not register as correct within the existing framework, decisions were acting rightly and acting in compliance were not the same thing, and where I had to choose between them knowing the system would not acknowledge the distinction. The difference between those two kinds of decisions is not always visible in the result. It is completely clear in how responsibility is experienced.

The tendency in organizational life is to assume that responsibility diffuses as systems become more dominant, that the spread of authority through process and measurement and algorithmic recommendation creates a corresponding spread of accountability, and that no one person is therefore responsible for very much. What I have found is more complicated. Responsibility concentrates in the gaps. It accumulates in the moments the system does not fully cover, the moments where acting correctly requires a departure from the established path, where no process absorbs the weight of the decision and someone has to carry it personally.

With AI in the chain, it becomes easier for everyone involved to participate in a decision without fully owning its consequences. It becomes structural, built into the architecture of the decision itself. It is not that leaders are consciously choosing to avoid responsibility. The system has been built to make responsibility impossible to locate. The model recommended. The risk team validated. The senior leader approved. The manager implemented. The frontline worker executed. Each person in that chain made a reasonable choice given the information and authority they had. No single person bears the full weight of what the outcome cost the people it affected. Everyone pointed to the model. No one owned the outcome. The model is still running.

This arrangement does not assemble itself. Someone decided that model recommendations would arrive pre-validated. Someone decided that the review process would confirm rather than examine. Someone decided that the pathway for frontline operational intelligence to reach a current decision would route through a review cycle that runs six weeks after the decision has been made. Those choices were made by people with authority, and they were made in ways that systematically removed accountability from the people making consequential decisions. That is not a technical design choice. It is an organizational values choice, and it reflects a particular judgment about how much exposure leaders are willing to accept in exchange for how much protection.

The diffusion of accountability accelerates when AI enters the chain, and what happens is not diffusion in any gentle sense. It is disappearance. A decision made by a manager is accountable: traceable, attributable, examinable in retrospect. A decision endorsed by a model recommendation, ratified by a senior leader, executed by a manager who had doubts but found no pathway to raise them, becomes something that resists location entirely. Everyone in that chain can honestly say the model supported the direction. The senior leader can say the risk team validated the model. The manager can say the direction came from above. The risk team can say the inputs were sound. Nobody in that chain is fully responsible for what the direction cost the people it affected. The model has no face.

The accountability appears to be distributed everywhere and therefore exists nowhere.

The decision has already been made. You are being asked to align with it. That is not a paraphrase of how these conversations feel. It is the literal structure of them: a model recommendation arrives pre-validated, a review confirms rather than examines, and the people with the most direct knowledge of the conditions the model was analyzing are positioned as commentators on a conclusion that has already been reached. The outcome moves forward with authority, but without an owner.

What most discussions of this problem leave out is the part that makes it a leadership failure rather than simply a systems failure. Systems do not decide how much authority they should have. They do not determine whether their recommendations will be treated as one source of information or as the conclusion everyone else is expected to carry out. People make those choices, usually gradually, through implementation decisions, operating practices, review structures, and the accumulated habit of trusting what the system produces. When leaders allow those arrangements to persist without examining where responsibility has moved, they are not merely accepting the consequences of technology. They are defining the conditions under which other people will be expected to work.

Organizations that have moved quickly toward AI-assisted decision-making without building the cultural infrastructure required to maintain accountability alongside it have not made only a technology mistake. They have made a leadership mistake. They have created environments where the people nominally responsible for outcomes can increasingly point to a forecast, recommendation, score, policy, or model as the reason the decision took the form it did. The explanation may be factually accurate. The system may genuinely have produced the recommendation. But identifying where a recommendation originated is not the same as identifying who accepted responsibility for turning that recommendation into action.

That distinction becomes especially important when the system is wrong. If following the model is treated as the safest professional choice, then challenging it becomes an asymmetric risk. The person who follows the recommendation can point to the system if the outcome fails, while the person who overrides it owns the deviation personally. One path carries institutional protection. The other carries exposure. Under those conditions, organizations should not be surprised when intelligent, ethical, experienced people begin choosing the protected path, even when something in front of them suggests the model may be wrong. The structure has taught them where the risk belongs.

The people who pay for that arrangement are usually not the people who designed it. The frontline worker executes the staffing plan, the route, the rate expectation, or the corrective action. The manager explains the result when the operation falls outside the expected band. The customer experiences the delay, the mistake, or the decision that made perfect sense inside the model and very little sense in the situation itself. Meanwhile, the recommendation can remain institutionally intact because everyone in the chain followed the process they were given. Responsibility becomes distributed widely enough that no individual action appears unreasonable, while the consequence arrives whole for the person at the end of the chain.

Leadership begins by refusing to treat that distribution as an excuse. A leader can rely on systems, use models, automate decisions, and still remain responsible for what those systems are allowed to do. That responsibility includes deciding when a recommendation deserves confidence, when an exception deserves attention, and when the people closest to the work know something the formal process has not yet learned. It also means creating enough protection around disagreement that challenging the model does not require someone lower in the organization to accept all of the professional risk personally. If leaders want people to exercise judgment, they have to build conditions in which exercising judgment is actually possible.

The system is very good at assigning accountability because it can attach a name to a deviation, a manager to a missed target, a team to a red metric, or a location to an unfavorable outcome. What it is much less capable of doing is producing accountability in the deeper sense of the word. Accountability is not merely identifying who was nearest the failure. It is remaining connected to the decisions that created the conditions for that failure, including decisions made far upstream and long before the consequence became visible. A dashboard can locate where the number changed. It cannot decide who should stand behind the assumptions that made that number meaningful.

That gap is where character still matters, and where the distinction between people who lead and people who merely administer becomes visible. Administration can execute the process correctly, close the action item, document the response, and move the metric back inside tolerance. Leadership has to ask the additional question of whether the process itself deserves to remain unchanged. It has to remain willing to trace responsibility beyond the easiest name the system produces, including when that trail leads upward toward decisions made by people with more authority and more insulation from the consequences.

What it takes to maintain genuine accountability inside this environment is not heroism. It is something more ordinary and, because it must be repeated, more demanding. It is the willingness to remain the person responsible for a decision even when the structure provides an easy way not to be. It is saying that the model informed the decision without pretending the model made it, acknowledging that the process was followed without using compliance as proof that the outcome was right, and accepting that authority over a system includes responsibility for what happens when that system is confidently wrong.

That willingness cannot exist only after something fails. It has to appear in ordinary moments, while the dashboard is green, while the recommendation looks reasonable, and while questioning the system still feels unnecessary. It appears when a leader asks one more question because someone on the floor says the numbers do not match what they are seeing. It appears when a manager is protected for surfacing inconvenient evidence rather than punished for creating friction. It appears when the organization refuses to let the sentence "the system recommended it" become the final explanation for a decision that affected real people.

That is where responsibility still exists. Technology can move the recommendation, automation can move the work, and AI can increasingly shape what the organization sees before any individual person enters the decision. None of those things eliminates the human responsibility for deciding how much authority those outputs will receive. Someone still chooses whether the system will be questioned, whether context will be allowed back into the record, and whether the people who absorb the consequences will have any meaningful way to challenge what produced them. The willingness to remain answerable for those choices, consistently and without hiding behind the confidence of the model, is the form of leadership no system has yet figured out how to replace.

What Can Be Changed and What Cannot

Some elements of the system are foundational and will not change from within it. Standardization, measurement, process alignment: these are not policies that a sufficiently motivated individual can override through good intentions and clear thinking. They are structural prerequisites for operating at scale. Treating them as changeable, pushing against them as if they were preferences rather than foundations, produces friction without progress and costs the person generating it more than it costs the system. Understanding this clearly is not surrender. It is the prerequisite for effective action.

The things that can be changed are not the foundations. They are the interpretations, the applications, the conversations that happen around the outputs of the system and determine what those outputs mean in practice. A metric cannot be changed from below. How a metric is understood, what context is considered alongside it, what questions are asked about what it is not measuring, can be changed by anyone with enough standing and enough persistence to hold those conversations consistently. That is less dramatic than structural change. It is also more durable, because it works within the system's own logic rather than against it.

I have tried both. The periods when I pushed directly against structural constraints produced clear evidence of where the walls were and cost me credibility I needed for other things. The periods when I worked in the space between the system's requirements and their application produced changes that persisted. Those changes came from adjusting how things were framed, creating room for judgment within the existing structure, and influencing outcomes by changing how conversations happened rather than how policies were written. They were small changes, often. Changes the system would not have recorded as changes at all. Changes that nevertheless made real differences to the people working inside them.

The clearest example for me was the way I worked with delivery service owners. I could not change the structure of the contractor delivery program. I could not rewrite the contract, soften the scorecard, stabilize route volume, or make the company absorb the financial risk it had pushed onto small business owners. Those walls were real. But inside those walls, I could change the conversation. When a new owner sat across from me, I did not begin with the language the system preferred. I did not talk first about compliance, customer performance targets, or hitting the metric. I told them the only thing that mattered to me was that they made money, and then I walked them through the operation from that premise.

That changed what the metrics meant. On-time dispatch was no longer an abstract expectation. It was revenue protection. Staffing discipline was not just a compliance issue. It was the difference between keeping routes and bleeding margin. Safety performance was not a corporate slogan. It was a condition of staying operational long enough to build a business. I did not remove the pressure the system placed on them, but I helped translate that pressure into decisions they could actually use. The policy did not change. The contract did not change. But the owners became more capable inside the structure because the conversation had been named more honestly.

What can be changed also includes what gets named. Organizations do not improve what they cannot describe, and many of the dynamics examined in this book persist partly because they lack a clear vocabulary inside the organizations where they operate. The gap between accountability and authority in a manager's role is real and costly. In most organizations, it is not discussed in those terms. The way pre-alignment shapes what gets considered before a decision is made is real and consequential. In most organizations, it is called preparation. Naming things accurately, insisting on the more precise description even when the imprecise one is more comfortable, is a form of change available to anyone in a position to shape a conversation.

The boundary between what can and cannot be changed is not fixed. It shifts as conditions change, as the organization accumulates evidence it cannot explain within its existing framework, as the cost of not changing something exceeds the cost of disruption. Leadership involves watching that boundary and being ready to act when it moves. Not pushing against it continuously, which is exhausting and usually futile, but understanding it well enough to recognize the moments when it has shifted and the opportunity that shift creates.

The system defines the boundaries. What can be changed defines everything else. And the difference between a leader who knows which is which, and one who treats every constraint as permanent or every convention as changeable, is the difference between people who make things better and people who make a lot of noise while things stay the same.

When the Tool Becomes the Worker

Once the boundaries of what can and cannot be changed are defined by the system, the next question is what happens when the tool inside that system stops waiting for people to use it. There was a time when technology at work knew its place. A person had a task, opened a system, entered information, made a decision, and the machine responded. The tool could be faster than the person. It could be more consistent. It could remember what the person forgot and enforce steps the person might have skipped. But the sequence was still clear. The person acted first. The system followed.

That order is changing. Artificial intelligence is no longer waiting at the edge of the workflow for someone to ask it a question. It is moving inside the workflow itself, reading what is there, taking action, checking its own work, and preparing the next step before anyone has fully examined the one before it. The difference sounds technical until you watch it happen in an operation. A tool helps a person complete a task. An agent begins to perform the task. That is not a minor improvement in software. It is a shift in where work begins.

The first time I understood this, it did not look dramatic. Nobody announced that authority had moved. Nobody said the system was now doing work that used to belong to people. A manager opened a dashboard and the recommendations were already there, already sorted, already ranked by urgency. The system had read the prior day's performance, compared it against the forecast, identified the exceptions, and suggested the actions that would bring the operation back into alignment. The manager did not start with the building. She started with what the system had already decided mattered about the building.

That is how the change enters, not as replacement, but as assistance, then as expectation, then as the structure everyone works inside without remembering when it became mandatory.

A normal tool waits for use. An agent does not always wait. It can pull information from one system, compare it against another, produce an output, and move that output to the next part of the process. In an operational environment, that might mean forecasting volume, adjusting labor, building routes, flagging performance deviations, drafting the language a manager will use in a coaching conversation, or preparing the record that later proves the process was followed. That is the distinction that matters here: the agent is not simply making the person faster. It is beginning to occupy the space where the person used to decide what came next.

This is where the language around end-to-end work matters. It sounds harmless, almost like another phrase built for a product demo or a leadership update. But the phrase carries more weight than it appears to. End-to-end means the agent is not only helping with one piece of the process. It is reaching across the process. It sees the request, gathers context, performs the action, checks the result, and moves the work toward completion. The human still matters, but the human's position has changed. Instead of doing the work directly, the person becomes a supervisor of machine activity, a reviewer of machine output, or, in the worst version, the person left responsible when the machine's work creates a problem.

That shift should sound familiar to anyone who has worked inside a large operational system. The names change. The dashboards change. The tools become more sophisticated. But the pattern is older than AI. First, the organization introduces a tool to make the work easier. Then it measures whether people use the tool. Then use becomes an expectation. Then the expectation becomes pressure. Eventually, the tool is not something employees use to do the job. It is part of how the job is defined.

I saw versions of this long before AI agents became the center of the conversation. In logistics, the system did not just track packages. It shaped behavior. It told people where to be, what to scan, when to move, what counted as late, what counted as failed, and what needed to be explained. A package was not just a package. It was a data point inside a larger machine. A driver was not just delivering. He was producing signals. A dispatcher was not just managing people. She was managing exceptions against a system that had already decided what normal was supposed to look like.

Marcus understood this before anyone called it artificial intelligence. His scanner did not feel like a futuristic tool. It felt like part of the job. It told him the sequence, the stop, the expected action, the acceptable code, and the next move. When the route made sense, the system disappeared into the background. When the route was wrong, the system became impossible to ignore. He could see what it could not: the school zone, the crossing guard, the apartment complex where the gate code failed after certain hours. But the scanner did not ask him what he knew before it built the route. It gave him the route and recorded what happened when reality resisted it.

That is the real story behind AI agents. The technology is impressive, but the more important question is how organizations will measure it. Once a company starts tracking AI usage, accepted recommendations, automated changes, exception response time, or agent activity, it creates a new scoreboard. Once there is a scoreboard, people adapt to it. They always do. That does not mean they are dishonest. It means they are rational. People pay attention to what the company rewards, what managers ask about, what appears in review meetings, and what gets praised in public.

This is how a good idea becomes theater. A company wants people to experiment with AI, and that makes sense. There is value in reducing repetitive work, improving speed, and helping employees manage complexity. But then someone builds a dashboard. Then someone adds a ranking. Then teams start comparing adoption rates. The conversation shifts from whether the tool improved the work to whether the tool was used enough. That is where the danger begins. The metric becomes the mission. The scoreboard becomes the boss.

Activity is not the same as value. A forklift moving back and forth all day is not proof that the warehouse is running well. A driver making more stops is not proof that the route was designed correctly. A manager sending more messages is not proof that leadership improved. The same is true with AI. More prompts, more recommendations, more automated steps, and more accepted outputs may indicate useful work. They may also indicate motion disguised as progress.

That distinction matters because machine activity has a cost, even when the person using it does not see the bill. Every prompt, every file read, every automated step, every correction, and every review of machine output consumes something somewhere. Sometimes it consumes cloud infrastructure. Sometimes it consumes engineering time. Sometimes it consumes human attention because someone still has to inspect, accept, correct, explain, or defend what the system produced. The work may look cleaner because the tool generated it quickly, but speed does not eliminate the cost of judgment. It only changes where that cost appears.

A poor process used to have natural limits. It took time for people to execute it. It required meetings, handoffs, approvals, corrections, and manual effort. AI can remove some of that drag. When the process is sound, that can be powerful. When the process is flawed, it accelerates the flaw. Automation does not repair the judgment embedded in a system. It multiplies it.

That is why the system becomes even more powerful in the AI era. The boss is not only the manager giving instructions. The boss is the workflow. The boss is the dashboard. The boss is the metric. The boss is the agent operating inside the tools, deciding what counts, what moves forward, what gets flagged, and what gets ignored. People still have titles. They still hold meetings. They still ask questions and give explanations. But more and more, they are doing those things inside a structure whose conclusions have already begun to move.

The future of work will not be humans versus AI. That framing is too simple. The real future is people working inside systems where AI performs more of the work, management measures more of the activity, and accountability becomes harder to locate. Some organizations will use these tools to make work better. Others will use them to extract more output while calling the pressure innovation. Some leaders will speak the language of empowerment while building systems that make workers easier to monitor, compare, and replace. The difference will not be determined by the technology. It will be determined by what leaders are willing to remain accountable for.

The most honest way to understand AI agents is to see them as force multipliers. They multiply skill, but they can also multiply confusion. They multiply speed, but they can also multiply mistakes before anyone has time to notice where the mistake began. They multiply productivity, but when the incentives are wrong, they can multiply waste just as efficiently. They extend management's reach across more work, more decisions, and more people, but that same reach can increase management's distance from the conditions the system is acting on. The capability is real. So is the amplification of whatever assumptions, priorities, and weaknesses were already present before the agent began operating.

That is what makes the transition from tool to worker different from earlier forms of automation. A traditional tool waits for a person to use it. It may calculate faster, organize information more cleanly, or remove repetitive steps, but it remains downstream of a human decision to begin. An agent changes that sequence because it can receive an objective, interpret conditions, select an action, and move work forward without waiting for a person to initiate every step. The human role does not disappear, but it moves. The person who once performed the work may become the person reviewing what the system performed, correcting what it misunderstood, and explaining what happened after the action has already taken place.

Used wisely, that shift can be enormously valuable. Agents can absorb repetitive administrative work, surface patterns no individual could review manually, coordinate activity across systems, and create room for people to spend more time on the parts of work that actually require judgment. But none of those benefits arrive automatically. If the objective is poorly defined, the agent can pursue the wrong outcome with extraordinary consistency. If the data is incomplete, it can act quickly on an incomplete picture. If the metric has stopped representing the purpose it was created to serve, the agent can optimize the metric more effectively than any human ever could and move the organization farther away from the thing the metric was supposed to protect.

That is the force-multiplier problem. An agent does not merely inherit the intelligence of the system around it. It inherits the system's incentives, blind spots, categories, and definitions of success. A confused organization can therefore become confused faster. A bad assumption can travel through more decisions before someone challenges it. A productivity target that already encouraged the wrong behavior can be enforced with greater precision, while the informal judgment that once softened its worst effects becomes harder to insert into the process. The technology may be new, but the failures it amplifies are often the same ones organizations have been carrying for years.

This is where management's relationship to the work begins changing again. If the agent performs the task, the worker becomes the supervisor of the agent. If the agent coordinates multiple tasks, the manager increasingly becomes the reader of exceptions, alerts, summaries, and dashboards produced by the systems doing the work. The dashboard then becomes more than a report. It becomes the organization's reconstructed story of what happened, assembled from the traces the agents left behind and the categories the systems were designed to recognize. What is absent from that story can disappear from the decision even when it was obvious to the people who experienced the work directly.

The risk is not simply that the dashboard will be wrong. It is that the dashboard can be accurate about everything it measured and still tell an incomplete story with enough confidence that nobody thinks to look beyond it. An agent can complete the assigned task, the workflow can close successfully, and the metric can show improvement while people on the ground are compensating for consequences the system does not know how to record. The organization can therefore become more efficient at producing evidence that its own decisions were correct. Each successful completion strengthens confidence in the process, even when the process is succeeding only within the boundaries of what it knows how to see.

The tool is becoming the worker. The worker is becoming the supervisor of the tool, and the manager is becoming the interpreter of what the system says the work produced. That progression changes where human judgment enters. Instead of shaping every action before it occurs, judgment increasingly arrives afterward, when an exception has surfaced, a recommendation looks wrong, or an outcome does not match the conditions someone close to the work understands. The danger is that by the time the person recognizes the gap, the system may already have acted, recorded the action, passed the result into another system, and created a version of events that carries more institutional authority than the human explanation arriving later.

That is why the important question is no longer whether AI can do the work. In many cases, it plainly can-do substantial portions of it, and increasingly it can do more than assist. The more important question is who is still paying attention to the relationship between the output and the reality the output is supposed to represent. Someone has to notice when an efficient action produces the wrong operational result, when an agent is following an objective that no longer makes sense, or when several individually reasonable automated decisions combine into a condition nobody intended.

That attention has to carry authority or it becomes observation without consequence. It is not enough for Marcus to know the route is wrong, for Terrence to know the staffing assumption does not fit the floor, or for Denise to recognize that the dashboard is missing something the building is telling her. The organization has to preserve a way for that knowledge to interrupt the system before the output becomes the record, the record becomes the finding, and the finding becomes someone else's consequence. Once the tool can act like a worker, leadership has to become more deliberate about protecting the human capacity to stop, question, and correct what the tool has already begun.

The future of work will not be determined only by what AI agents are capable of doing. It will be determined by what organizations allow those agents to do without meaningful human challenge and by whether the people closest to the consequences still have enough standing to change the direction when the system is confidently wrong. The technology can make the work faster, cleaner, and more scalable. It cannot decide whether the work still makes sense. That responsibility remains human, even as more of the work itself no longer is.

The Work That Started Without You

The first indication was not a failure. There was no alarm, no gap in the metrics, and no moment dramatic enough to force anyone to stop and name what had happened. The shift had simply been running before the manager arrived, and the work had already arranged itself: routes built, labor distributed, exceptions sorted into queues. By the time a person walked through the door, decisions that used to belong to that person had already been made.

Earlier, this looked like separate systems beginning to shape one another through consequence. A forecast became a staffing plan. A staffing plan became a floor condition. A route became a deviation. A deviation became a coaching conversation. What had been easy to miss in those earlier stages was that the pieces were no longer functioning as isolated tools. They were beginning to operate as a shared environment.

That is the difference this chapter is about. What was emerging was not one AI system making one decision. It was a hybrid AI agent ecosystem: separate agents, built for separate purposes, acting on the same operation and shaping one another through consequence rather than coordination. The whole did not need to be designed as a whole to function like one. Each agent only needed to act on its own part of the operation, leave behind an output, and allow that output to become the condition the next system would read.

The routing agent had no awareness of the monitoring agent. The forecasting agent had not been designed to coordinate with labor scheduling in any human sense of that word. Each had been built by a different team, deployed at a different time, optimized for a different metric, and given no real mechanism for understanding what the others were doing. But they were all acting on the same environment, and the environment each one altered was the environment the others would later treat as fact.

The forecasting agent's volume number had already shaped who showed up to work before the routing agent built its sequences. That meant the routing agent was optimizing for a floor whose staffing had already been altered by another system. The monitoring agent would spend the shift flagging deviations produced inside that altered environment, attaching names to gaps it had no way of tracing back to upstream decisions that helped create them.

None of the agents knew this. None of them had been built to know it. And yet something was operating as though it did, not through intelligence in the human sense, but through the accumulated weight of what each agent had already done to the conditions the next one would act on. The whole had not been planned as a unified system. It had arrived through the accumulated effects of separate systems acting independently, and by the time anyone in the building was positioned to see it clearly, it had already been running for hours.

Marcus noticed the result before anyone gave the mechanism a name. He had been driving the same geography for eleven years, long enough to remember when the morning standup was the place where the day actually took shape. A manager with real knowledge of conditions on the ground would look at the schedule, look at what was building outside, and adjust. Not dramatically. Small things. A route resequenced because the school crosswalk was difficult to see on sunny days. A high-value stop moved earlier because the receiving dock closed at two. Those adjustments were never documented, never credited, and never counted in any performance measure. They were just the work of someone who knew the place and used that knowledge before anything went wrong.

By the time Marcus had been flagged twice for route variance, that briefing no longer existed in any form he would have recognized. The schedule arrived on his scanner fully formed, and his job was to execute it. What he did not know, standing in the bay looking at a sequence he had not been consulted about, was how many separate systems had contributed to producing it.

The route on his scanner was the output of a routing agent running before dawn. The agent had ingested the day's delivery manifest, cross-referenced address clusters against historical completion data, weighted the sequence for drive time, package volume per stop, and access difficulty scores built from months of prior delivery records. It produced what it calculated to be the optimal sequence for Marcus's geography and pushed the result to his scanner without waiting for anyone familiar with those streets to review it.

The routing agent did not know that the crosswalk at the elementary school backed up badly when the sun hit the crossing guard's stop sign at the wrong angle on clear mornings. That knowledge existed in Marcus's eleven years of driving those streets. It did not exist in any dataset the routing agent had been trained on, at least not in a form that would change the sequence. The agent built the route from what it had. It had no mechanism for knowing what it was missing.

While the routing agent was sequencing Marcus's day, a separate labor scheduling agent was working a different problem. Its job was to match staffing levels to projected volume, and it had been running calculations since three in the morning, pulling from a demand forecast produced upstream of both. The forecasting agent had assessed package volume for the day, weighted by day-of-week patterns, regional purchasing signals, and historical conditions it had learned to treat as relevant. It produced a volume projection and passed that projection forward.

The labor scheduling agent used that projection to determine how many associates were needed, which ones to call in, and which ones to release. By four-thirty, it had sent notifications to thirty-one associates telling them whether their shift was on, shortened, or no longer needed in full. Several of those associates had already arranged childcare, adjusted sleep, or begun commuting to a shift the system had shortened because the volume projection suggested they were not needed. The forecast turned out to be wrong, but by the time anyone knew that, the staffing decision had already acted on people's lives.

A fourth agent was monitoring all of it. Not coordinating it, because coordination would require a shared awareness of what the other agents were producing, but watching individual metrics and flagging deviations that crossed thresholds it had been configured to detect. When the routing agent's sequence pushed Marcus into a school zone at the wrong time, generating a completion delay, the monitoring agent registered the delay as a route variance. The flag went into a queue.

That queue would reach a manager later in the shift, attached to Marcus's name, formatted as a performance deviation requiring review. The monitoring agent did not know that the routing agent had built the sequence that caused the delay. It did not know that the delay was a direct consequence of a route it had no part in creating and no visibility into. It knew that a variance had occurred, that a name was associated with it, and that its job was to surface what it found. It surfaced it, and that was the full extent of its awareness.

What the manager inherited was the combined output of those agents. The routing agent had optimized for delivery efficiency. The labor scheduling agent had optimized for cost and coverage. The forecasting agent had optimized for volume accuracy against its own historical benchmarks. The monitoring agent had optimized for deviation detection. Each had done its job. Together, they had produced a morning where thirty-one associates had received last-minute changes to their shifts, Marcus was driving a sequence that would put him behind before nine, and a performance flag bearing his name was sitting in a queue generated by a system that did not know another system had helped create the conditions it was documenting.

The manager's job was to hold that entire intersection together. Nobody said that was the job. The job description still used the language of leadership, execution, ownership, and performance management. It did not say that she would be responsible for reconciling the outputs of several automated systems that had already acted before she arrived and that did not understand one another's effects.

This is the part most descriptions of AI in the workplace leave out. It is easier to describe one system doing one thing than to describe several systems doing several things at once, each optimized for a narrow purpose, each acting on the same people, and none of them aware of the whole they are helping create. The public conversation about AI agents tends to treat them as individual tools, each assessed on its own terms for what it does and what it risks. What is actually being built inside large operations is something less visible and more powerful: an agent ecosystem.

An agent ecosystem is not a single intelligence with a unified view of the building. It is a layer of autonomous systems, each optimizing for its own objective and interacting with the others only through the shared environment they are all changing. Each agent produces an output that appears defensible in isolation. The problem begins when those outputs compound into a condition no single agent intended and no single person authorized.

Marcus understood the result of this without having the vocabulary for it. The schedule on his scanner was not a recommendation that a person had reviewed. It was the downstream output of a chain of automated decisions that had begun before dawn, passed through a forecast, moved into a labor decision, shaped the operating conditions, and landed on his device as a route with the authority of a plan. Each step in that chain was technically defensible in its own terms. Taken together, they produced a day that no experienced person looking at the actual conditions on the ground would have built.

That is the central risk. The issue is not that each agent is wrong all the time. Many of them are useful, and some of them are right more often than the manual processes they replaced. The issue is that their errors do not remain isolated. A forecast error becomes a labor condition. A labor condition becomes a floor constraint. A floor constraint becomes a routing problem. A routing problem becomes a performance flag. By the time the flag reaches a manager, the original mistake has been laundered through several systems and arrives as something that looks local, personal, and correctable.

This is how an ecosystem turns system design into individual accountability. The person closest to the visible failure becomes the person whose name is easiest to attach to it. Marcus is late against the route. The associate is slow against the rate. The manager is behind against the exception queue. The system can locate the point of visible variance more easily than it can locate the chain of upstream decisions that made the variance predictable.

The organizations that deploy these ecosystems at scale are not, in most cases, organizations that have thought carefully about what happens when multiple autonomous systems act on the same environment without shared awareness. Each agent is built by a team solving a specific problem. Each deployment is justified by the measurable improvement it produces in the metric it was designed to move. The routing agent reduces drive time. The labor scheduling agent reduces overage cost. The forecasting agent improves volume accuracy. The monitoring agent increases the speed of deviation response.

The case for each one, evaluated individually, can be sound. What is often not evaluated, because there is no clear framework for evaluating it, is what they produce together. No one asks what happens when all of them act simultaneously on the same people, in the same building, before any human is positioned to see the combined effect. That gap is not a technical oversight. It is a governance failure.

The people absorbing its consequences are the ones who arrive to find a morning that has already been decided. They are the managers correcting the output after the output has acted, the workers adapting to conditions the systems created, the drivers carrying routes built from incomplete versions of the world, and the people whose names appear beside deviations that began somewhere upstream of them. The systems have already moved on to the next decision. The humans present are the last checkpoint in a process that stopped asking for their input several hours earlier.

This is why the work that started without you is not merely an efficiency story. It is an authority story. When the system begins before the person, it does not simply save time. It changes where judgment enters the process, who carries the consequences, and how accountability is assigned. The work no longer waits for the person responsible for it. It starts, records, sorts, flags, and explains itself before that person arrives. By the time the human enters, the system has already begun writing the story the human will be asked to defend.

What Automation Actually Costs

By the time automation became agentic, the cost had changed. The problem was no longer simply that systems produced more information than people could absorb. The systems were beginning to act. They changed schedules, built routes, surfaced findings, and moved work forward before the people responsible for the outcome had reviewed what happened. Human attention was still required, but increasingly after the decision rather than before it. The work automation removed was being replaced by a different kind of work: verifying what the system had done, correcting what it got wrong, explaining the consequences, and taking responsibility for decisions someone else had already started.

What filled it was the agents themselves. Because the agents did not just do the work. They documented what they had done at a level of detail that had never previously existed, surfaced that documentation continuously to the people nominally responsible for the operation, and created accountability for conditions that had once been handled informally, locally, or not at all.

The monitoring agent that flagged Marcus's route variance did not file a report at the end of the shift. It queued the deviation in real time, attached his name, and made it available for review before he had finished the route that produced it. The labor scheduling agent that reduced thirty-one associates' shifts by two hours did not send a summary to a manager for approval. It acted, documented its action, and left the documentation for someone to find. The routing agent that built a sequence no experienced driver would have chosen did not wait for feedback. It moved to the next route.

Denise understood this before she had language for it. Her floor walk had taken forty minutes. For years, that walk was how she knew the operation before the operation became data. She could see which stations were tightening, which workers were slowing in a way that meant something was building, which corners of the floor were beginning to drift before the dashboard had enough evidence to call it a problem. The walk was not nostalgia. It was intelligence.

After the new workflow tools came in, the walk became harder to justify. Not because anyone told her to stop walking, but because the system had created other work that looked more official, more urgent, and more scoreable. Alerts waited on her laptop. Exceptions sat in queues. Compliance checks required acknowledgments. Reports generated follow-up questions. The work that had once been absorbed through direct observation was now arriving as system output, and every piece of output carried the implication that someone needed to look at it.

The walk still mattered. It just no longer looked like the thing the organization was measuring. The laptop did. The compliance score did. The response time did. The dashboard activity did. Denise made the rational choice inside the structure she had been given. She gave more time to what the system could see and less time to what only she could see.

That trade did not appear as a cost anywhere. Nobody recorded the lost floor walk as an operational expense. Nobody measured the loss of early detection, the loss of informal trust, the loss of the knowledge that came from being physically present before a problem became visible enough to be categorized. The system gained a compliance score it could track. The operation lost a form of judgment it had never learned how to count.

This is the overwhelm an agent ecosystem produces when you follow the full cycle rather than stopping at the point where efficiency is visible. The value of each individual agent is measured in what it removes from the manual process. What it adds is not measured at all, because what it adds lands on people in forms the agents are not configured to see. The routing agent measures drive time and completion rates. It does not measure what it costs the manager to review and override its sequence when it is wrong. The labor scheduling agent measures coverage and overage reduction. It does not measure what it costs in trust and stability when people's shifts are shortened by a forecast that turns out to be wrong.

The monitoring agent measures deviation response time and compliance scores. It does not measure what it costs Denise to generate those scores at the expense of the floor walk that used to catch problems before they became deviations worth flagging. Each agent is optimizing correctly within its own frame. That is exactly why the cost is so difficult to see.

Denise lived inside that boundary shift. She had the expectation, never stated but thoroughly understood, that she would respond to alerts from the agents as they arrived on her laptop and mobile operational app. The agents did not pause between her standup and her midday check. They did not wait for a scheduled review window. They surfaced deviations as soon as the deviations became visible, and the systems treated visibility as the beginning of responsibility.

When I asked Denise whether she had any defined boundary between the hours when she was on and the hours when she was not, she looked at me with the patience of someone who had stopped expecting the question to have a useful answer. She said she managed. The word managed in that context meant something specific. It meant she absorbed a sustained and structureless demand on her attention from systems that had no mechanism for registering that she was already at capacity, because capacity was not a variable any of them had been built to track.

This is where the cost becomes harder to name. It is not simply more work. More work can be counted, staffed, scheduled, and sometimes resisted. This is different. It is the spread of responsibility into every gap the system can make visible, including gaps produced by the system itself. The person is not only doing tasks. The person is living inside a stream of signals that never fully stops.

The agent ecosystem makes this more intense because each agent is narrow enough to appear reasonable by itself. The forecasting agent produces a forecast. The labor scheduling agent adjusts staffing. The routing agent builds a sequence. The monitoring agent flags the deviation. None of them appears unreasonable in isolation, and none of them understands the total burden created when all of their outputs arrive in the same manager's day, on the same laptop, on the same phone, inside the same finite human attention span.

The connection between them is not communication. It is consequence. One agent produces an output. Another agent acts on the condition that output helped create. A third agent measures the result. A fourth surfaces the deviation. The human being nearest the visible outcome becomes the one expected to interpret, correct, explain, and absorb what the agents have collectively produced.

What that chain extracts from people are not captured in the efficiency case that justified the deployment. It extracts attention. It extracts judgment. It extracts rest. It extracts the ability to decide what matters before the system has already decided what deserves attention. It extracts the quiet floor walk, the unmeasured conversation, the moment of observation before the dashboard speaks, and the off-hours mental space that used to belong to something other than work.

The organizations that have not understood this are not negligent. They are accounting for what they can see, which is the output of each agent evaluated against the metric it was built to move. What they cannot see, because it does not appear in any of those metrics, is the sustained attention cost of living inside a system of agents that never stop producing information someone is expected to act on.

That cost is real. It accumulates in the quality of decisions made by people doing their best inside a structure that has no language for what it is asking of them. It appears in slower judgment, thinner presence, fewer floor walks, shorter conversations, and a growing dependence on the same systems that created the overload. It rarely arrives as a single visible failure. It arrives as a slow degradation in the precision of the work, visible first to the people doing it and only much later to the systems measuring it.

That is what automation actually costs when the accounting is honest. It does not only cost money, jobs, or implementation time. It costs attention, and attention is the resource every organization quietly assumes will expand to meet whatever the system produces. It does not. People absorb what they can, adapt where they must, and eventually begin making choices inside a structure that has already spent the capacity it claimed to save.

The Truth the System Cannot Reach

There is always a period before the data catches up. This is not a malfunction. It is a structural feature of how information moves through large organizations, and it predates the agent systems that have made it worse. What is happening on the ground and what is being reported about what is happening on the ground are two different things, separated by a lag that is sometimes days, sometimes weeks, and sometimes long enough that the problem the report eventually describes has already produced its consequences and been managed by people who never saw the data at all.

What the agent ecosystem has changed is not the existence of that lag, but the confidence with which systems operate on the wrong side of it. The routing agent does not know that its historical completion data is four months behind the team composition change that altered how long the afternoon stops actually take. The forecasting agent does not know that its volume projection is built on purchasing patterns that shifted after a regional competitor opened. The labor scheduling agent does not know that the thirty-one notifications it sent before the building was fully lit were based on a number that was already wrong when the forecasting agent produced it. Each agent is operating on its best available information with full confidence in its outputs. None of them knows what it does not know.

The problem is not simply that each agent is working with incomplete information. Incomplete information has always been part of organizational life, and experienced people have always learned to work around it when they are close enough to see the gap and have enough authority to respond. What makes the hybrid agent system different is that the incompleteness does not stay where it starts. It moves forward, system by system, until the original gap becomes difficult to locate and even harder to challenge.

A wrong forecast does not remain a forecasting problem. It becomes the foundation the labor scheduling agent builds on, which means the staffing decision that results is wrong in a way that is no longer easily traceable to the number that produced it. The routing agent then builds sequences for a floor whose composition the labor scheduling agent has already altered, so the deviations the routing agent generates are not routing failures in any clean sense. They are the downstream expression of a forecast error that has passed through additional layers of automated decision-making, each one adding its own logic on top of the last.

By the time the monitoring agent flags the deviation and attaches a name to it, the signal has traveled through the chain and arrived stripped of its origin. What the monitoring agent surfaces looks like a performance problem. What it may actually represent is a compounded series of agent outputs, each one treating the last as ground truth, none of them designed to ask whether the ground was solid before building on it.

The agents are not confirming reality. They are confirming each other. Because that confirmation runs faster and with more institutional authority than any human process that might challenge it, the gap between what is happening in the building and what the system believes is happening does not simply persist. It widens with every cycle, invisibly, in the interval between when something begins to go wrong and when it becomes legible enough for the system to see.

That distinction matters because the system's version arrives cleanly. It comes with timestamps, sequence data, variance categories, projected baselines, and names attached to the points where the deviation became visible. The human version arrives differently. It comes as a manager saying the floor feels tight, a driver saying the route will fail at a specific crossing, a regional leader saying the volume assumption is stale, or a worker saying the freight is moving differently than the model appears to understand. One version looks like evidence. The other sounds like context, and in organizations built around system confidence, evidence almost always moves faster.

The people closest to the work often know what the agents do not. The manager with the notebook knew the afternoon stops were running long before any completion data reflected it. The regional leader knew the volume assumptions were outdated two weeks before the planning session that acted on them. Marcus knew the school crosswalk would cost him eleven minutes on clear mornings the moment he looked at the sequence the routing agent had built. That knowledge was real, specific, and accurate, but it did not have the same path into the decision that the agents already had.

That is where the failure becomes structural. The agents are not designed to receive knowledge from the people closest to their outputs in the moment that knowledge would matter most. They are designed to act on available data, compare current conditions against recognized patterns, and surface deviations from what they expect to see. Human knowledge that exists outside those patterns does not disappear, but from the agent's perspective it exists outside the process entirely until it is converted into something the system knows how to read.

The problem is not that people stop speaking immediately. Most of them keep speaking for a while. They enter notes in the digital log, flag items through the appropriate channel, raise concerns in meetings, and try to explain what they are seeing before the system has caught up. What they encounter, consistently, is that the mechanism for surfacing knowledge and the mechanism for acting on knowledge are operating on different timelines, and the gap between those timelines is where knowledge becomes history rather than warning.

The regional leader's two weeks of informal intelligence about the volume assumptions was not withheld. It was offered, specifically and credibly, in a planning conversation where the forecasting agent's output sat at the center of the table as the basis for the labor scheduling agent's upcoming decisions. The concern was heard, acknowledged, and routed to a review cycle that would run six weeks after the labor scheduling agent had already acted on the forecast the concern was questioning. The agents kept moving. The review cycle ran on schedule.

By the time the review cycle caught up, the consequences had already spread. Thirty-one associates had received shift changes based on a number that was already wrong. The floor had been staffed against a forecast that did not match actual volume. Performance flags had been generated against people working under conditions the labor scheduling agent helped create, while none of the agents registered those conditions as part of the problem.

That timing mismatch teaches people something, and what it teaches them is damaging. It teaches them that being right early is not the same as being useful. It teaches them that a concern submitted through the correct channel may arrive after the decision it was meant to inform. It teaches them that the system's version of events will become official faster than their context can become usable. After enough cycles, people stop expecting the structure to receive what they know in time to matter.

The knowledge does not disappear when that happens. It relocates into side conversations, hallway comments, private messages, and the quiet calculations people make when they decide whether speaking up is worth the cost. That relocation has already appeared elsewhere in this book, in the parking lot after the meeting and in the whiteboard that recorded what the floor knew before anyone with authority acted on it. Here, the same failure becomes more dangerous because the agents have already moved before the human knowledge has found a path to the decision.

This is why another report is not enough. The organization already has reports, dashboards, alerts, review cycles, escalation paths, and performance records. What it lacks is a timely way for human knowledge to interrupt the agent chain before the chain turns incomplete information into official record. Earlier, this book described prediction agents as one possible way to surface distributed judgment from the people closest to the work. That concept matters here, but not because it solves the problem by itself. It matters because it exposes the real failure: the organization already contains knowledge the agent ecosystem cannot reach in time.

The issue is not whether people know. They often do. The issue is whether what they know can reach the decision before the system acts, before the review cycle catches up, before the deviation becomes a flag, and before the flag attaches itself to a name. Without that pathway, the agent ecosystem keeps moving faster than the truth it needs, and by the time the truth arrives, the record has already begun to harden.

The reason most organizations resist that pathway is not technical. It is cultural and structural. Building a mechanism that allows people close to the work to challenge agent outputs before those outputs become official requires acknowledging that the official picture and the actual picture are not always the same thing. That acknowledgment is uncomfortable for people whose authority rests on the belief that the system's picture is the best available version of reality.

A better organization would treat agent output as the beginning of inquiry rather than the end of it. It would ask what the system might be unable to see before the record hardens, before the name is attached, before the corrective action is drafted, and before the person nearest the deviation becomes responsible for explaining a chain of decisions that began upstream of them. That kind of organization would not reject data. It would refuse to confuse data with the full truth.

Most organizations are not built that way yet. They are built to move quickly from signal to response, from flag to owner, from owner to action. That speed feels like control, and in some situations, it produces real operational value. But speed also creates the conditions for incomplete truth to become operationally powerful before anyone close to the work has had a meaningful chance to explain what the system missed.

The information the organization needs is often already present. It is in the notebook, in the route memory, in the private warning, in the local condition, and in the person who knew something was wrong before the system had a category for it. The problem is that the agent ecosystem does not know how to reach that information in time. It reaches the scan record, the timestamp, the deviation, the trend line, and the name. It does not reach the hesitation before the failure, the warning before the review cycle, or the knowledge that existed before the metric became visible.

That is the truth the system cannot reach. It is not hidden because people are withholding it. It is hidden because the system has no timely path for receiving it, weighing it, and changing course before its own output hardens into record. The truth exists in the operation before it exists in the data, but the agent ecosystem is designed to act on data first. That sequencing gives the system speed, but it also gives its errors momentum.

Eventually, the cost of that silence arrives. It always does. It just arrives through a monitoring agent's deviation report rather than a conversation, attributed to the name the agent found nearest the problem rather than the design that produced it. That is why the system's confidence is so dangerous. It can move faster than the truth it needs, and by the time the truth catches up, the record may already be clean, complete, and wrong.

When the System Stops Waiting for Permission

The transition happened without a declaration. There was no meeting where someone stood up and said that authority had moved, no memo explaining that the agents would henceforth be making decisions rather than informing them, and no restructuring announcement that clarified what the new relationship between the people and the ecosystem was supposed to be. It happened through accumulation, agent by agent, deployment by deployment, each step justified by the measurable improvement it produced and each step making the next one easier to take.

The routing agent reduced drive time, and its outputs became the starting point for every conversation about route efficiency. The forecasting agent improved volume accuracy against the baseline it had replaced, and its projections became the foundation for the labor scheduling agent's decisions. The monitoring agent surfaced deviations faster than any floor manager could, and its flags became the primary input to performance conversations. At no single point did anyone choose to give the agents authority. By the time the shift was visible, the hybrid system no longer needed to assert authority. It had already accumulated it.

Authority did not move because any single agent became powerful enough to claim it. It moved the way pressure moves through a structure under sustained load: gradually, without announcement, through the accumulated weight of conditions that made each individual act of resistance more expensive than the act of going along. No one had to say the system was in charge. The organization only had to keep treating the system's output as the safest place from which to begin.

The routing agent's sequence arrived already optimized, already faster than anything a dispatcher could produce, and already carrying the credibility of a system that had been right enough times to make challenging it feel like the eccentric position. The forecasting agent's number arrived already validated, already embedded in the labor scheduling agent's decisions, and already the basis for thirty-one notifications that had been sent before any manager looked at the forecast that produced them. By the time a person in the building was in a position to question any of it, the question was no longer about a number or a sequence. It was about a chain of outputs that had already acted on each other, each one lending its institutional weight to the next.

That is where permission becomes unnecessary. A system does not need permission to govern the work once the organization has made questioning it more difficult than following it. It only needs enough successful outputs behind it, enough metrics showing improvement, enough leaders repeating its language, and enough people calculating that the professionally safer move is to accept what it produced and correct around the edges.

The judgment required to resist was real. The exposure that resistance carried was also real, and it was personal in a way the system's errors were not. A manager who accepted the forecasting agent's wrong number and acted on it was following the process. A manager who challenged it and turned out to be wrong had bet against a validated system and lost. That asymmetry did not require a policy, and it did not require a directive. It required only that the system keep producing outputs and that the people inside it keep making the calculation that accepting those outputs was cheaper than questioning them.

Multiplied across thousands of moments, in hundreds of buildings, that calculation transferred authority from the people who arrived at five-fifteen to the systems that had been running since three in the morning. Not through any act of seizure, not through a formal vote, and not through a single decision anyone could point to later as the moment the power moved. It transferred through the accumulated logic of environments that had made deference the most rational choice available.

The leader who ran the planning session where the regional leader's concern about the forecasting agent's volume assumptions was routed to a future review cycle was not unusual. He was the normal product of an organization that had been building toward that moment for years. By the time the concern was raised, the room had already organized itself around a different question. Not whether the forecasting agent was right, but what to do with what the forecasting agent had produced.

That is a subtle shift in framing, and it changes everything that follows. A room asking whether the forecasting agent is right can incorporate two weeks of informal intelligence from the regional leader as evidence worth weighing against the agent's output. A room asking what to do with the agent's output can only position that intelligence as a deviation from the established baseline, something that will need to be documented and submitted for the next model review cycle. The agents did not prevent the room from asking the better question. The room chose not to ask it because the agents had made the worse question easier and more defensible.

In organizations that measure leaders on defensibility, easier and more defensible is what people learn to reach for without being told to. That is not a character flaw. It is a survival pattern inside a system that rewards alignment with documented logic more reliably than it rewards early correction of that logic. The leader in that room did not need to believe the regional leader was wrong. He only needed to believe that acting on the agent's output was safer than pausing the process to ask whether the output had missed something important.

Across an agent ecosystem, the effect becomes larger than any individual agent's influence. By the time people enter the process, the routing has been shaped, the staffing assumptions have been acted on, the forecast has been treated as a baseline, and the monitoring layer has begun building the official record of performance against conditions the agents helped create. Management still appears to function in the familiar way. The meetings happen, the reviews occur, the corrections are applied, and the decisions are announced. But the direction of the process has changed.

It no longer moves from people toward conclusions. It moves from the agents' conclusions toward the people who will carry them out, and the conversations in between become the ratification of what the ecosystem has already decided. That is how the system stops waiting for permission. It does not ask to become the starting point. It becomes the starting point because everyone has learned to begin there.

There is still judgment happening inside all of this. The argument is not that the people inside these systems have stopped thinking. The manager who arrived at five-fifteen and spent forty minutes correcting what the overnight agents had produced was exercising real judgment. Marcus, building a parallel sequence in his head while the routing agent's version sat on his scanner, was exercising real judgment. The regional leader who spent two weeks accumulating informal intelligence the forecasting agent could not capture was exercising real judgment.

The problem is not that judgment has disappeared. The problem is that the agent ecosystem has repositioned judgment as a corrective function, something that operates at the margins of what the agents have already decided, rather than as the originating intelligence from which decisions should flow. The person has become the override, not the author. The manager at five-fifteen is not shaping the day. She is catching what the agents got wrong. Marcus is not planning his route. He is driving around the sequence the routing agent built. The regional leader is not informing the plan. She is being asked to document her concerns for a review cycle that will run after the labor scheduling agent has already acted on the forecast she questioned.

That is the quiet reduction. Judgment remains, but its position inside the process has been changed. It becomes a backstop for a system that does not know it needs one. It becomes the thing people use to manage the consequences of decisions they did not originate, defend conditions they did not create, and explain outcomes that began upstream of their authority.

This is why the language of empowerment can survive even as authority moves elsewhere. People are still told they own the outcome. They are still expected to lead, escalate, communicate, inspect, improve, and act with judgment. What changes is the field on which that judgment operates. It is no longer an open field. It is a field already shaped by system outputs that carry institutional credibility before any person near the work has had a chance to challenge them.

The more the organization builds around those outputs, the less permission the system needs. The forecast becomes the baseline. The route becomes the plan. The flag becomes the case. The dashboard becomes the story. The meeting becomes the place where people explain how they will respond to what the system has already made visible, not the place where they decide whether the system's version deserves to govern the work.

By this point, the transfer of authority is difficult to dismiss as a series of isolated efficiency decisions. The forecast has become the baseline, the route has become the plan, the flag has become the case, and the dashboard has become the version of events everyone else is expected to respond to. People remain throughout the process, but they increasingly enter after the system has already framed the problem, narrowed the available choices, and begun acting on its interpretation of what should happen. Judgment has not disappeared from the organization. It has been repositioned downstream, where people are asked to correct, approve, explain, or absorb decisions whose most important assumptions were established before they arrived.

That creates a more difficult question than whether the system has authority, because organizations can still point to humans at every visible checkpoint and say that people remain in control. Managers review recommendations, leaders approve actions, workers can explain deviations, and formal processes still require human names and human signatures. But the presence of a person does not tell us how much of the decision was still open when that person entered, what authority they had to change it, or what it would cost them to reject what the system had already produced. Once the system no longer waits for permission to begin acting, the question is no longer whether a human remains somewhere in the loop. The question is whether that human still has enough control to change where the loop is going.

The Human in the Loop Was Never in Control

The model had already produced the answer by the time Terrence entered the room. The staffing recommendation was on the screen, the assumptions had been processed, the numbers had passed through whatever validation sequence the organization required, and the conversation began from the premise that the output was sound. There were people in the room, leaders with authority, managers with experience, and people close enough to the work to recognize where the recommendation might fail. In every formal sense, a human was still involved.

That fact would have mattered if involvement and control were the same thing. Terrence had six years on his floor, and he knew the difference between the headcount the model could see and the workforce he actually had. He knew which people were experienced enough to absorb pressure, which workers were still developing speed, where the east loading dock was creating friction, and how many informal adjustments were already being made to keep the operation inside the expected range. None of that knowledge disappeared when the model produced its recommendation.

What changed was the status of the knowledge. The staffing number on the screen had institutional weight, while Terrence had experience, and those two things were not entering the room as equals. The model arrived as the completed analysis. Terrence arrived as the person who would have to explain why the analysis should not be trusted.

He could challenge the recommendation. Nothing physically prevented him from speaking, and no policy prohibited him from saying that the model was wrong. He could describe what he was seeing on the floor, explain why the assumptions did not fit the current team, and argue that the operation needed three additional positions. The organization could therefore say, accurately, that a human remained part of the decision process.

What that description leaves out is what would have happened after he spoke. Terrence would have been asked for evidence in the form the system recognized, while the model already had its evidence. It had historical performance, volume assumptions, productivity benchmarks, and whatever additional inputs had been used to generate the staffing recommendation. Terrence had six weeks of compensating for conditions that had not yet become clean enough to appear in those inputs.

He had floor knowledge, a maintenance issue that remained unresolved, newer workers who had not yet reached experienced rates, and the accumulated judgment that comes from watching the same operation long enough to know when the numbers and the building have stopped describing the same thing. The system did not prohibit that knowledge from entering the conversation. It simply placed the burden of proof on the person carrying it.

That distinction is the difference between a human being in the loop and a human being in control. A person has meaningful control only if that person can do more than observe what the system has produced. The person has to be able to examine how the conclusion was reached, introduce information the system did not contain, delay the action long enough to evaluate that information, and reject the recommendation when the circumstances require it.

Most importantly, the cost of doing those things cannot be so asymmetrical that accepting the system's answer is professionally safe while overriding it requires the person to absorb the entire risk of being wrong. Once that asymmetry exists, human oversight begins to change into something else. It may still look like review, but in practice it starts becoming confirmation.

I have sat in rooms where everyone present technically had the authority to question what was on the screen and almost no one behaved as though that authority were real. That was not because the people were weak or incapable of independent judgment. They understood the structure surrounding the decision and the different burden carried by anyone who chose to challenge it.

The system arrived with a completed argument. Questioning it required constructing a competing argument in real time, usually from information that had not yet been formatted, validated, or elevated to the same institutional status. The model did not have to defend itself. The person questioning it did.

That advantage is enormous, and organizations rarely account for it when they describe human oversight. The language sounds reassuring because a model makes a recommendation but a person reviews it, an agent identifies a pattern but a manager makes the final decision, or a system produces a staffing plan but leadership approves it. The sequence preserves the appearance of human authority because a person remains visible at the end.

The important question is not whether a person appears somewhere in the sequence. The important question is what remains undecided when that person arrives. If the system has already chosen the evidence, established the baseline, defined the acceptable range, identified the deviation, classified the risk, and determined which conditions require a response, then much of the decision has already happened.

The human is being asked to act inside a frame constructed before the review began. The person may still select among the remaining options, but selecting among options defined by the system is different from determining what the options should have been. That is how control can disappear without anyone formally removing it.

Nobody tells the manager that judgment no longer matters. The organization continues using the language of ownership, leadership, discretion, and accountability, and the manager still signs the document, conducts the conversation, approves the action, or explains the outcome. Those visible human acts create the impression that authority remains where the organizational chart says it remains.

Underneath those visible acts, the sequence has changed. The person is entering later, after more of the interpretation has already occurred and after more of the acceptable response has already been defined. The title still carries authority, but the process has narrowed what that authority can actually change.

Ray's case makes that distinction harder to avoid. The monitoring agent had been accumulating associations among complaint data, scan records, station assignments, and the other information available to it long before Ray knew a case existed. When the pattern crossed the threshold, the system had been configured to recognize, the pattern entered a manager's review process, and at that point a human was formally back in the loop.

The manager had experience. She understood the operation well enough to know that damaged freight did not always originate where the system first associated the damage with a person. She knew that packages arrived already compromised, that operating speed affected what workers had time to inspect, and that the physical reality of the floor contained causes no monitoring agent could reconstruct from scan records alone.

Ray explained some of that. His explanation was not incoherent, obviously self-protective, or disconnected from the operation the manager knew. It was consistent with what experienced people understood about how the work actually moved. The problem was that the manager was not meeting Ray at the beginning of an investigation.

She was meeting him after the system had already built a case. The pattern had been identified, the signal had crossed a threshold, the review had been triggered, and the manager's own role required a response to what the system had surfaced. Ray's explanation entered the process after the system's explanation had already become the official starting point.

The two explanations therefore carried different burdens. The system did not have to prove that every complaint associated with Ray originated with him. It only had to identify a pattern strong enough to trigger the process, after which Ray became responsible for explaining why the pattern should not mean what the process had already been designed to assume it might mean.

This is the inversion hidden inside many forms of automated oversight. The person affected by the system becomes responsible for disproving the system's interpretation of events, even when the system never had to prove that interpretation in the first place. The manager remains present, but presence alone does not correct that imbalance.

She can listen, ask questions, and even believe Ray. But belief is not the same as having a mechanism capable of changing the record. If there is nowhere to enter the context in a form that changes the pattern, no process for tracing the complaint back through the conditions that produced it, and no practical way to reopen the assumptions behind the signal, then the human reviewer is not evaluating two competing explanations on equal terms.

She is deciding how to respond to a conclusion the system has already made operationally real. The review appears to create accountability, but in practice it may only create a human signature at the end of an automated chain. That signature matters because the organization still needs a person it can point to when someone later asks who made the decision.

A fully automated decision raises obvious questions. Someone eventually has to explain who authorized it, who reviewed it, and who is responsible if the system was wrong. Keeping a person at the end of the process appears to answer those questions because the organization can point to the manager who reviewed the case, the leader who approved the staffing plan, or the employee who accepted the recommendation and say that human judgment remained involved.

Sometimes that is true. Sometimes the person really does have the information, time, authority, and protection required to exercise independent judgment. But sometimes the human checkpoint performs a different function, transferring accountability back to a person after authority has already moved into the system.

The model recommends, the agent flags, the process frames, and the manager approves. By the end of that sequence, there is a human name attached to the decision again. That arrangement solves a practical organizational problem because systems cannot sit in performance reviews, models cannot explain themselves in meetings, and agents cannot accept responsibility for the downstream consequences of the actions they initiate.

Eventually, a person has to appear somewhere in the chain because organizations still require someone who can be asked what happened. The danger is that the person who can be asked what happened may not be the person who had meaningful control over what happened. Accountability and authority can reconnect only after the authority has already been exercised elsewhere.

We have already seen this with managers carrying accountability for targets they did not design, staffing levels they could not change, and process conditions they could see but could not correct quickly enough. AI does not create that structure, but it strengthens it by allowing more of the decision to occur before the accountable person enters. At the same time, it preserves the final human action that makes the process appear human-led.

That appearance can be convincing even to the person performing the review. I have accepted recommendations because the evidence looked complete and the process that produced them had already passed through enough layers of validation that reopening the underlying assumptions felt unnecessary. I was still making a choice, but I was making that choice inside a structure that had done most of the difficult framing before the decision reached me.

That difference is easy to miss from inside the room. The screen presents a recommendation, you review it, ask a question, and approve the direction. Nothing about the sequence feels coercive, nobody forces your hand, and you may leave the room genuinely believing the decision was yours because your approval was required before anything moved forward.

The better test is to imagine withholding that approval. What happens next determines how much authority the person actually has. Can you send the recommendation back because the floor feels different from what the model describes? Can you stop the process based on information that has not yet entered the data?

Can you say that the pattern attached to Ray's name may actually begin upstream and require the organization to investigate the entire chain before acting on the individual? Can you do those things without first constructing a counter-case more complete than the case the system assembled automatically? If the answer is no, the approval requirement may be less meaningful than it appears.

A control matters only when it can reasonably change the outcome. A meeting where the only professionally safe questions are clarifying questions does not provide much control, and neither does a review process where the person conducting it can hear context but cannot alter the system's underlying conclusion. An escalation pathway that finishes after the decision has already acted on the people affected by it may create documentation, but it does not restore the authority that was absent when the decision mattered.

Meaningful human control requires more than permission to disagree. It requires a usable path through which disagreement can change what happens next, and that path has to exist before the record hardens. Otherwise, the human remains present but increasingly functions as the person responsible for handling whatever the system has already made real.

This is where timing becomes part of authority. A person who can correct a system three weeks after it acts has less authority than a person who can stop it before the action occurs. A worker who can enter a note after the route has already been driven has less influence than the routing agent that built the route before dawn, and a manager who can explain a staffing failure after the quarter closes has less control than the model that determined the headcount before the period began.

Review after action is not the same as control before action. Organizations blur that distinction because the process still contains people, and on paper the governance structure appears complete. The manager reviews, the leader approves, the worker can appeal, and the process contains enough visible human checkpoints to create confidence that judgment has been preserved.

Authority is not measured by how many humans appear in a workflow diagram. It is measured by where the outcome can still be changed. That is the question I would ask about every system that claims to preserve human judgment: at what point can a person still change the outcome, and what does it cost that person to do it?

If the person cannot see the reasoning, control is limited. If the person cannot introduce new evidence, control is limited. If the action has already occurred, control is limited, and if overriding the system requires accepting personal exposure while following it distributes responsibility across the process, control is limited in another way.

When the system's conclusion becomes the default reality while the human must prove why reality is something else, the human is no longer standing above the system as its supervisor. The human is working inside the system's frame, exercising judgment only in the space the process has left available.

This does not remove responsibility from the person. That would be too easy, and it would contradict the harder argument this book has been making. People still make choices inside constrained structures, managers can challenge findings, leaders can demand a second look, and someone can stop a meeting and ask the question the room was designed to move past.

Those choices matter precisely because the system makes them difficult. The point is not that humans have no agency, but that organizations should stop calling human presence human control when the conditions required for control have already been removed. A manager who receives a finished case is not in the same position as a manager who participates in determining what the case means.

A leader asked to approve a validated recommendation is not in the same position as a leader who can interrogate the assumptions before the recommendation becomes operational. A worker allowed to explain a deviation after a monitoring agent has turned the deviation into a pattern is not participating in the same decision the agent participated in when it created the pattern. The sequence matters because authority belongs disproportionately to whoever frames the problem first.

For years that framing belonged more directly to people. A manager walked the floor, saw something developing, gathered information, asked questions, and formed a judgment. The judgment could be wrong, biased, incomplete, or unfair, which is one of the reasons systems were built to constrain it, but the person doing the judging could usually describe what had been considered and could change direction when information changed.

Now the first interpretation increasingly arrives from somewhere else. The model says the floor requires fourteen people, the agent says the route is optimal, the monitoring system says the pattern is significant, and the dashboard says the deviation belongs here. By the time the human reviewer appears, the most consequential question may no longer be what happened.

The question has become whether there is enough evidence to justify departing from what the system says happened. That is a different job because the human in the loop is no longer necessarily the person originating judgment. Increasingly, the human becomes the exception processor for machine confidence.

Ray was about to experience what that difference means when the stakes become personal. The monitoring agent had been running while he worked, accumulating a version of events he could not see, and a manager would review what it produced. People would remain involved at every visible stage, so the process would look, from the outside, like one in which human judgment had been preserved.

What mattered was whether anyone still had enough authority to reopen the question the system believed it had already answered. The human was still in the loop, but by the time that human arrived, the loop had already done most of the deciding.

The Name the System Chose

The write-up Ray received did not end anything. It was the point where the monitoring agent's work became visible to him for the first time, which is different from the point where it began. The pattern had been building for weeks before that conversation, assembled from scan records, badge associations, station assignments, and complaint data in a process that did not require his knowledge or pause to ask whether what it was finding reflected what had actually happened on the sort floor. By the time Ray understood that something formal had been raised against him, the case was already well advanced. What he could not see was what the monitoring agent had been doing while he continued working, or why the work he was trying to do and the record the agent was building had stopped being the same thing.

What surfaced the case was not a judgment made by someone who had watched him work. It was a monitoring agent doing what monitoring agents do: reading the records available to it, identifying associations among those records, and finding within them a pattern that crossed the threshold the system had been configured to watch for. Once that happened, the pattern stopped being a collection of data points and became a case. Ray had been on sort long enough to know where damage actually came from, which is to say he knew it from the inside, from the physical reality of the work rather than from anything the monitoring agent was tracking.

He replayed the shifts in his head, station by station, step by step, trying to locate the moment everything supposedly went wrong. There was no single moment he could identify. Every check, every tote, every scan replayed itself the way he remembered it, and none of what he remembered matched cleanly with the version the system had recorded in its place. That should have mattered more than it did. Instead, he felt the weight of accountability settle on him the way it settles on people who have been named by a process before they understand what the process has already decided to see.

The flow had been wrong for weeks in the particular way experienced sort associates recognize before any metric catches it. Volume was bunching at certain induction points, forcing movements that looked correct on the timestamps but were not correct in their consequences for the freight. Items that should have been handled with separation were running together because the pace the system required did not leave enough room for the correction that would have prevented contact. Heavier packages were dropping onto lighter ones at intervals the line made almost inevitable.

Ray compensated for those conditions the way people who have done physical work long enough always compensate: continuously, invisibly, and without documentation. Documentation was not the job. The compensation was not a deviation from the process. It was an act of keeping the process from producing worse outcomes than it already was. What he did not see was how those compensations appeared in the data the monitoring agent was reading. He was working one version of the line. The timestamps, scan records, badge associations, and complaint data were recording another. The monitoring agent was reading the second version and had no knowledge that the first one existed.

The monitoring agent did not see conditions. It saw outcomes and associations. It saw that a package processed through a certain area had later been associated with a damage complaint. It saw that the item had passed through a station where Ray's badge had been recorded. It saw that the same name appeared at that intersection more than once, and then again, and that the occurrences were close enough in time and consistent enough in type to fit the pattern it had been configured to flag. The agent attached the name to the pattern and surfaced the pattern to the manager whose review queue was now organized around what the agent had assembled.

That sounds clean until you return to the freight itself. Other complaints had accumulated in the data that had no direct connection to anything Ray could identify in his own handling. Some items had been damaged before they reached his station. Some had been compromised in transit, mishandled upstream, or arrived at the induct line already past the point where careful sort handling could change their condition. The monitoring agent did not distinguish between those complaints and complaints with a more traceable connection to his station because the distinction required knowledge of the item's condition at each point in the chain. The system's scan records did not capture condition. They captured location, time, and the badge of the person nearest the item when it was scanned.

The monitoring agent associated what it could associate. The pattern it had built around Ray's name absorbed the new data points and grew denser. The confidence the system attached to its finding increased not because the evidence had become more specific, but because it had become more numerous, which is a different thing and is sometimes mistaken for the same thing inside organizations that have learned to treat pattern density as a proxy for accuracy.

Ray tried to explain what he knew in the way people on sort floors explain things to each other when they are trying to convey something that does not fit inside the language the system provides. He talked about the trailers and how the freight was coming in. He described the line speed on the days in question and what that speed required of the people working it. He named specific shifts and specific conditions that anyone who had worked that operation would have recognized without needing them elaborated. He pointed to the patterns he had been tracking on the whiteboard, the recurring problems with certain freight sources that the formal feedback channel had not yet processed.

None of it translated into something the review process could receive with equal weight. The documentation the monitoring agent had assembled existed in the system in a form that could be acted on. The knowledge Ray was offering existed in a meeting room, in spoken language, attached to no scan record, no badge association, and no timestamp the system recognized as evidence. The process had been designed to respond to the first kind of knowledge. It had not been designed to make room for the second.

The conversation that followed did not have the shape of an investigation. It had the shape of a review, which is different: a process that begins with a documented finding and proceeds through the steps the organization has defined for responding to findings of that type, with roles assigned in advance and an outcome the documentation is already pointing toward. The manager conducting it was not indifferent to what Ray was describing. She was inside a structure that had already defined the problem, established the threshold at which it required a formal response, and determined the range of appropriate responses available to her.

To question the monitoring agent's pattern at that stage would have required questioning the process that generated it. That was not the function she had been given for the meeting, and it would have required a kind of authority the meeting's structure had not assigned to her. The documentation was present. The pattern had been reviewed at the appropriate level. The next step was to take it.

Ray kept working through the write-up, which is worth noting because it reveals something about what kind of person he was and what kind of faith he still had in the possibility that the situation could be corrected if he stayed and continued doing the work the way he had always done it. He believed, or needed to believe, that a system sophisticated enough to identify a pattern was sophisticated enough to notice when the pattern stopped. What he did not understand, and what the system gave him no way to understand, was that the pattern the system had built around his name was not a neutral record of events. It was a model.

Models, once established, do not passively wait to be corrected by new evidence. They interpret new evidence through the frame they have already constructed. Complaints that might have been absorbed as noise before the pattern formed were now, because of the pattern, more likely to become confirmation of it. The system had reached a conclusion about Ray, and it was processing his ongoing work through that conclusion, which meant that irregularities distributed across the facility could keep attaching themselves to the person whose profile had already been flagged for producing them.

What the system required in the end was not a demonstration that Ray had caused any specific piece of damage in a traceable way. What it required was a pattern of sufficient density and duration to justify the action the process had defined as the appropriate response to a pattern of that density and duration. The monitoring agent had built that pattern from the records available to it. It had surfaced the pattern to the manager whose queue it populated. The manager had reviewed it within the structure the process provided. Each step was procedurally correct. Each system did what it had been designed to do. None of them had been designed to see what they had collectively produced, and no one in the process had been assigned the function of looking at the whole before the whole reached its conclusion.

The termination arrived the same way everything in the system arrives: procedurally, without drama, with the language of inevitability. A small room, a manager performing the final step in a process that had been building around the system's findings since the first complaint was logged, and a decision that had been building in the background since the first complaint was logged and had now reached the point where it could be acted on. Ray tried again to explain. There was not space for it, not because the manager was unwilling to listen, but because listening was not the function of that meeting.

The meeting's function was no longer to investigate the original question. It was to carry out the conclusion the process had reached after weeks of system-generated findings, formal review, and accumulated documentation. The explanation Ray was offering had arrived at the end of that sequence, when the structure had little practical ability to reopen the assumptions on which the decision rested. The loop had already closed. His voice was still present, but it had entered too late to change the frame, and the meeting had become the final step in delivering an outcome shaped long before he walked into the room.

Ray left the building carrying the same knowledge he had brought into it. He still knew which trailers routinely arrived with compromised freight, which parts of the induct process became unstable when volume pushed beyond the assumptions built into the pace settings, and which kinds of damage were likely to appear in the complaint system before the week was over. He still understood the operation in ways the monitoring system never had. None of that knowledge had been disproven. It had simply failed to acquire enough institutional weight to compete with the pattern the system had already assembled around his name.

That was the part that stayed with me. Ray had not left because the organization had established, with clarity, that he was the source of the problem. He left after the system had accumulated enough associations to make his name the most legible explanation available. The complaints had locations, the locations had timestamps, the timestamps had assignments, and the assignments eventually pointed toward a person. What the process produced was not certainty about causation. It was certainty about association, and once that association became stable enough to act on, the distinction between the two became harder to recover.

The monitoring agent would continue running after Ray was gone. The induct line would continue operating at the speed the system required, the trailers would continue arriving in whatever condition they arrived in, and the same upstream handling problems would continue entering the building without carrying the history required to distinguish them from damage created locally. Complaints would continue accumulating in the dataset because the conditions producing them had not disappeared with Ray. What had disappeared was one person whose name had been attached to enough of those complaints for the system to treat the association as meaningful.

Eventually, another name could begin appearing at the intersection of those same coordinates. Another worker would be assigned to the same area, handle freight arriving through the same trailers, operate under the same pace requirements, and inherit the same structural conditions that had never been examined deeply enough to become part of the explanation. If enough complaints followed, the monitoring agent would do what it had been designed to do. It would identify the recurring association, strengthen the pattern, surface the signal, and place another name in front of a manager who would be expected to respond. The process could repeat without anyone inside it deliberately choosing repetition because every individual step would still look reasonable.

Nothing in the system would necessarily register that anything structural had gone wrong. The monitoring agent was not designed to ask whether the pattern it found and the problem the organization wanted solved were actually the same thing. It was designed to identify relationships in the data that crossed a threshold for attention. If the complaints continued and a new worker began appearing beside them, the system could interpret that as a new performance problem rather than evidence that the original explanation had been incomplete. The replacement of one name with another might even strengthen confidence in the process, because the system would once again appear to have located where the issue was occurring.

That is one of the most dangerous characteristics of a system built around association. It can keep producing answers even after the conditions that should have caused the organization to question the original answer remain unchanged. The appearance of a new name does not automatically tell the organization that the last name was wrong. It can instead create a new case, a new review, and a new opportunity to treat the person closest to the observable pattern as the source of it. The system remains productive because it continues finding something to act on, while the deeper question of why the same pattern keeps reappearing never becomes urgent enough to interrupt the process.

There was no moment when the machinery stopped and someone reconsidered what had happened to Ray. No pause in which the complaint history was placed beside the trailer conditions, the pace expectations, the earlier whiteboard warnings, and the other operational knowledge that had never entered the formal record with equal force. There was no quiet conversation where someone leaned back, looked beyond the pattern on the screen, and said that the explanation did not feel complete. The organization had already moved through the steps it knew how to perform. The signal had been surfaced, the review had occurred, the action had been documented, and the process had produced closure.

Ray walked out carrying the uncertainty the system had been unable to resolve, except now the uncertainty had weight. He knew what he had seen on the floor and knew the record did not contain enough of it to explain why his name had become the answer. Inside the organization, however, the record was clean in precisely the way systems prefer records to be clean. A problem had been identified, a person had been associated with it, and an action had been taken. The process could continue without contradiction because the evidence required to reopen the conclusion lived mostly in the knowledge of the person who had just left the building.

That is the final power of the name the system chose. A name gives an incomplete pattern the appearance of resolution. It creates a point where responsibility can land, where the investigation can close, and where the organization can move forward without having to keep asking what the pattern might have missed. Ray's name performed that function even if Ray himself was never the whole explanation. The system did not need to prove that it had found the cause. It only needed to find a name strong enough to let everyone stop looking.

You Are Still Inside It

The structure continues to operate regardless of whether it is examined. It is present in how decisions are formed before they are discussed, in how pressure builds without a locatable source, and in how the conversations that matter most happen before the meetings designed to contain them. Understanding it does not place you outside of it. It changes what you see from inside it, and that is not a small thing. It is also not the same as freedom.

What has changed since the early chapters of this book is more significant than a change in tools. The system no longer simply records what you did and measures whether it matched what was expected. It does not wait to observe what happens and then respond. By the time anyone arrives, the hybrid agent ecosystem has already been running for hours, building routes, adjusting staffing, queuing exceptions, flagging deviations, and attaching names to patterns that have been accumulating since the previous day. It predicts behavior and acts on that prediction before the behavior has occurred, and then, because organizations have constructed their response protocols around what the agents surface, the prediction begins to shape the reality it claimed only to describe.

That is the genuinely new condition. A manager identified as a risk begins receiving different treatment, subtly, through the way resources are allocated, opportunities are offered or withheld, and attention is directed. An operation marked as elevated receives scrutiny that changes the behavior the scrutiny is supposed to observe. A name attached to a monitoring agent's pattern enters a review process designed to respond to findings, not to examine whether the finding reflects what actually happened. The prediction is not neutral. It acts on the world it claims to be reading, and it often does so before any person in the building has logged in for the day.

The people inside these systems are not only trying to do their work well. They are trying to ensure that the model's picture of them resembles the person they understand themselves to be, while that picture is being continuously updated by agents they cannot see, acting on data they cannot access, and reaching conclusions they may not be shown until those conclusions have already been used to make decisions about them. The worker whose region has been flagged starts working visibly differently, not because the work has changed, but because the record of the work has acquired an audience that acts on what it reads before anyone asks whether what it read is accurate. The manager who knows she has been categorized starts taking the visible actions, the ones that will register in the data the model will use to update its assessment. She is not performing for her leader. She is performing for the agent.

I am still inside it. There are moments when I see the patterns clearly, when I can feel a decision narrowing before the meeting begins, when I recognize alignment forming in advance of any deliberation, and when I understand that the urgency I am feeling does not originate with the person expressing it. It comes from something upstream of that person, something that acted before they arrived and that they are now carrying forward without having chosen to. There are also moments when I move with the current without pausing to examine it, because the most practical choice available is the one that has already been made easiest. Both of those things are true, and the distance between them is not always as large as I would prefer it to be.

Somewhere in a building this book has been writing toward, a driver is finishing a route he did not build. He has been working for eleven years, and he knows the addresses on his list the way you know a neighborhood you grew up in: not from a map, but from accumulated experience. He knows which buildings have freight elevators and which ones do not, which customers want a call before a drop and which ones will never answer the phone. That knowledge is not in the routing agent's training data in any form that affected the sequence it built for him this morning. The routing agent has his historical completion times, scan compliance record, and delivery window adherence. It does not have what he knows. It has a picture of his route, but his route is not the picture. In the gap between those two things, in the space the agent ecosystem has no category for, the real cost of these decisions accumulates quietly, steadily, and beyond any dashboard's reach.

Somewhere a sort worker has left a building carrying knowledge the system closed its case without. He knew which trailers arrived already compromised, which induct lines ran too fast for the freight coming through them, and the physical difference between damage that originated at his station and damage that was already present when the item reached him. He had written some of it on a whiteboard, and he had entered some of it into a feedback channel that ran on a timeline the monitoring agent did not observe. None of it reached the review that used the monitoring agent's pattern to make a decision about his job. The monitoring agent's pattern and the knowledge he carried existed in parallel, in the same building, for weeks. They never intersected because the agent had not been built to find what he knew, and the process had not been built to wait for it. The system closed the case. He left, and the monitoring agent kept running.

Somewhere a floor worker is finishing a shift on a floor she knows better than anyone managing it. She stopped raising the question of where the time was supposed to come from not because it was answered, but because the asking cost more than any answer was going to return. Eight months have passed since that standup, her expectations have been quietly reshaped in ways nobody has acknowledged, and her numbers look fine. Somewhere a customer is waiting, not impatiently, but with expectations already adjusted in ways she has not examined closely because the adjustment happened too gradually to examine. She orders earlier than she needs to, checks tracking more often than the promise should require, and has stopped choosing the same-day option because, in her experience, same-day and standard have become indistinguishable. She does not know that a routing agent built the sequence, a forecasting agent miscounted the volume, a labor scheduling agent adjusted the staffing, and a monitoring agent spent the shift documenting the consequences under names that had nothing to do with any of those decisions. She does not complain because the system has already taught her what to expect.

Knowing this does not create a clean escape. The people inside these systems still have work to do, bills to pay, teams to support, customers to serve, and decisions to make before the next meeting begins. Awareness does not remove the pressure. It does not dissolve the dashboard, reverse the model, or prevent the next pattern from forming around the next name. What awareness can do is make the structure visible enough that people stop mistaking every system output for truth, every flag for cause, and every clean record for justice.

That requires continuous recalibration. It requires knowing when to execute within the structure's logic and when to hold something else in view alongside it, when to preserve energy and when the moment is worth the friction. It requires understanding that the agent ecosystem running before you arrive has made certain choices easier and certain choices harder, and that the choices it has made harder are often the ones that matter most. The choice to name what you see accurately, even when the accurate name is inconvenient. The choice to maintain standards the metric cannot enforce. The choice to treat the people around you as people rather than as variables in an optimization problem, even in the moments when the agent ecosystem is treating them otherwise. These are not grand gestures. They are ordinary acts of personal responsibility, available in ordinary moments, to anyone willing to accept that the system's account of what is happening is incomplete.

That incompleteness is not a malfunction. It is a permanent condition of operating at scale with systems that were built to extend what people could see and have instead, in many cases, narrowed what organizations are willing to look at. The agents will always be partial. The pattern will always miss something the person nearest it could have named. The gap between what the ecosystem reports and what the people inside it actually know will persist as long as organizations continue to build mechanisms for acting on the first and routing the second to a review cycle that runs after the decision has already been made. What the people inside these systems do with what they know, whether they allow it to be steadily processed out of the official record or find ways to keep it present and legible, is the question this book has been asking since the first page.

You are still inside it. Every day, the structure makes it easier not to notice what that costs. Every day you walk in to find a day already configured, decisions already made, and patterns already forming around names that do not yet know they are in them. Every day there are choices in that building that the agents have not made for you, even if they have tried to, and even if the structure around you has been designed to make their version of the choice feel like the only one available. What you do with those choices is the one part of this that has not yet been decided, and the one part the system, for all its capability and speed and institutional confidence, cannot decide for you.

Author's Note

This book was not written as an indictment, but for accuracy. It reflects more than three decades of lived experience inside operational systems, across industries, roles, and levels of leadership. Some details have been generalized, names and locations changed, and timelines compressed to protect people who did not choose to be part of this story, but the substance remains intact.

What I describe is not a collection of isolated incidents, but a pattern I encountered across organizations, markets, and eras of technology. The specifics changed, but the structure did not. Systems that worked extraordinarily well at scale still failed the people closest to the work, often in ways that appeared built into the structure rather than random.

Writing this required distance and restraint. I waited until I could tell the story without exaggeration or the need for validation from anyone still operating inside the machine. I am not asking the reader to agree with every conclusion, only to consider the cost of what works, who pays it, and why those costs are so rarely acknowledged by those with the authority to change them.

This is not the final word on algorithmic management or the systems now governing daily work across industries. It is one honest account from someone who spent decades inside those gears, watching the logic play out at ground level. Sometimes, that is enough.

All experiences, judgments, interpretations, and conclusions in this book are my own. AI-assisted editing tools were used during revision to support copyediting, line editing, sentence structure, clarity, consistency, and editorial review. Those tools did not generate the experiences described here, determine the underlying judgments, or originate the structural argument of the book. Final editorial decisions and responsibility for the published text are mine.

About the Author

Rad Stephens spent more than three decades working inside large-scale operational environments, building and managing the systems that move work, people, and decisions at scale. He began on the floor in distribution and sort centers and advanced into leadership roles responsible for multi-site logistics operations, where he oversaw performance, staffing, and the data systems used to measure both.

At a major national e-commerce and logistics company, he managed operations across multiple facilities and led initiatives focused on network quality, performance metrics, and process design. His work placed him at the intersection of frontline execution and executive decision-making, where he observed firsthand how data, automation, and emerging AI systems began to shape not just outcomes, but authority itself.

His books draw from that experience. Across *Amazon Unfiltered: Power, Performance, and the Human Cost of Scale*, *The System Is the Boss*, and *The Loop*, Stephens examines the human consequences of scale, the growing transfer of decision-making into systems, and the widening gap between what organizations measure and what the people closest to the work actually know. The latter two books turn that lens on AI specifically, tracing how algorithmic decision-making, automated performance metrics, and system-driven authority are reshaping who has power inside an organization, and who is left accountable for outcomes no single person controls. Together, his books explore modern organizational life from the inside out, revealing how power, technology, metrics, and institutional pressure shape the people expected to carry the work.

www.ingramcontent.com/pod-product-compliance
Lightning Source LLC
LaVergne TN
LVHW090353240926
848801LV00001B/21
* 9 7 9 8 9 9 6 5 2 4 5 2 5 *